T0330977

Kaizen

How to Successfully Transition
into a Lean Organization

Kaizen
How to Successfully Transition into a Lean Organization

Marc Helmold
IU Internationale Hochschule Berlin, Germany

World Scientific

NEW JERSEY · LONDON · SINGAPORE · BEIJING · SHANGHAI · HONG KONG · TAIPEI · CHENNAI · TOKYO

Published by

World Scientific Publishing Co. Pte. Ltd.

5 Toh Tuck Link, Singapore 596224

USA office: 27 Warren Street, Suite 401-402, Hackensack, NJ 07601

UK office: 57 Shelton Street, Covent Garden, London WC2H 9HE

Library of Congress Control Number: 2024030242

British Library Cataloguing-in-Publication Data
A catalogue record for this book is available from the British Library.

KAIZEN
How to Successfully Transition into a Lean Organization

ISBN 978-981-12-9243-9 (hardcover)
ISBN 978-981-12-9280-4 (paperback)
ISBN 978-981-12-9244-6 (ebook for institutions)
ISBN 978-981-12-9245-3 (ebook for individuals)

For any available supplementary material, please visit
https://www.worldscientific.com/worldscibooks/10.1142/13821#t=suppl

Desk Editors: Aanand Jayaraman/Lum Pui Yee

Typeset by Stallion Press
Email: enquiries@stallionpress.com

Printed in Singapore

This book is dedicated to my wonderful wife Takako,
my daughters Ayumi and Manami.

Preface

Lean management is a modern concept to eliminate waste and thus concentrate on customer satisfaction. The concept of lean management integrates significant factors like modern leadership, agile attributes, and continuous improvement. It helps companies achieve a sustainable competitive advantage.

The book explains the lean management concepts, including Kaizen, Kata, and Keiretsu, in a way that enterprises can gradually implement them to achieve a long-term competitive advantage.

The book provides a holistic and practical approach to lean management throughout the business value chain. The lean management framework and tools demonstrate the optimal design and use of methods, tools, and principles for companies and organizations. The author describes comprehensively how lean management enables companies to concentrate on value-adding activities and processes to achieve a long-term, sustainable competitive advantage. A wealth of best practices, industry examples, and case studies are used to reveal the diversity and opportunities of lean management methodologies, methods, and principles.

About the Author

Marc Helmold (MBA) is a full-time professor at IU Internationale Hochschule in Berlin, Germany. He teaches bachelor, master, and MBA courses in business management, operations management, lean management, quality management, and leadership. He has previously held several executive leadership positions at OEMs in the automotive, railway, and aviation industries. At IU Internationale Hochschule, where he has been working since 2016, he also conducts workshops on lean and quality management and leadership.

Contents

List of Acronyms and Abbreviations

AI	Artificial Intelligence
DIN	German Industry Norm
EN	European Norm
ERP	Enterprise Resource Planning
ISO	International Standardization Organization
GoGA	Go Gemba Audit
HA	Combining Theory and Practice
LAW	Lean Audit Workshop
PDCA	Plan, Do, Check, Act
QMS	Quality Management System
RI	Mastery
Shu	Discipline
SIT	Short Interval Technology
VSA	Value Stream Audit
3D	Dreidimensional
3G	Gemba, Genbutsu, Genjitsu
5G	Gemba, Genbutsu, Genjitsu, Gensoku, Genri
5S	Sort, Set in Order, Shine, Standardize, Sustain

List of Figures

List of Tables

Chapter 1

Lean Management and Kaizen

1.1 Scope and Definition

Lean management is a modern, agile concept for process optimization for manufacturing and service companies to implement throughout the internal and external value chains (Helmold & Terry, 2021). Lean management focuses on identifying inefficiencies (waste), making these inefficiencies transparent, and transforming them into value-adding activities (Ohno, 1990). Waste can be divided into obvious and hidden waste. The value chain reaches from the suppliers (upstream) towards a company's operation and the customers (downstream), as illustrated in Figure 1.1 (Slack & Brandon-Jones, 2021). Inefficiencies or waste can refer to a process, a product, a procedure, or an area along the company's value chain for which the customers are not willing to pay. The customer is the central point in the lean management concept. The primary objective of the lean management concept is to create value for the customer through the optimization of processes and resources. Hence, it is possible to steadily optimize processes and workflows along the value chain in order to satisfy customers. Lean management seeks to eliminate any waste in terms of transportation, inventory, motion, overproduction, overprocessing, or defects. Lean management focuses on identifying weaknesses in each step of a business process and then transforming or eliminating those steps that do not create value (Bertagnolli, 2020). The philosophy has its origins in Japan in the field of operations but is presently widely implemented

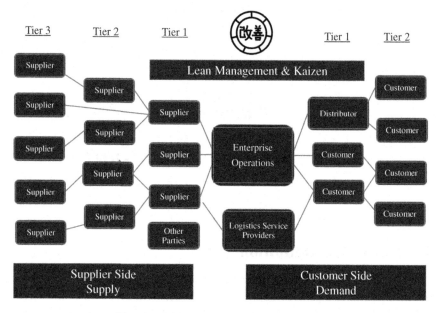

Figure 1.1.   Lean management and Kaizen.

across the world in different industries (Belekoukias *et al.*, 2014). Lean management focuses on the following:

- Making the customer the focus of the operation.
- Defining value and value addition from the standpoint of the end customer.
- Eliminating all waste in all areas of the value chain.
- Continuously improving all activities, processes, purposes, and people.
- Making the people the centerpoint of value-adding services and processes.

Holistic lean management implementation can lead to significant productivity improvements of more than 100%, as previous studies have shown (Helmold *et al.*, 2022). In some areas where the lean management concept is new, companies have processes that consist of more than 90% waste (Helmold & Terry, 2021). Lean management facilitates modern and agile leadership characteristics and must be part of a cultural transformation of an enterprise.

Lean management is closely associated with Kaizen, or the philosophy of continuous improvement. Continuous improvement ensures that every employee contributes to the improvement process within the value chain. The management method acts as a guide to building a successful and solid organization that is constantly progressing, identifying real problems, and resolving them. Lean management is based on the Toyota production system (TPS), which was established in the late 1940s. Toyota put into practice the five principles of lean management with the goal of decreasing the number of processes that were not producing value; this became known as the "Toyota Way." By implementing the five principles, Toyota found that significant improvements were made in efficiency, productivity, cost efficiency, and cycle time. Lean management incorporates the following five guiding principles that are used by managers within an organization as guidelines to the lean methodology (Helmold & Terry, 2021):

(1) Identify value in all processes of the value chain.
(2) Conduct value stream mapping.
(3) Create a continuous workflow.
(4) Establish a pull system in which the customers are the focus.
(5) Facilitate a culture of continuous improvement.

Identifying value, the first step in lean management, means finding the problem that the customer needs solved and making the product the solution. Specifically, the product must be part of the solution that the customer will readily pay for. Any process or activity that does not add value, importance, or worth to the final product — meaning it does not add usefulness and the customer is not willing to pay for it — is considered a waste and should be eliminated (Liker, 2004). Value stream mapping refers to the process of mapping out the company's workflow, including all actions and people who contribute to the process of creating and delivering the end product to the consumer. Value stream mapping helps managers visualize which processes are led by what teams and identify the people responsible for measuring, evaluating, and improving the process. This visualization helps managers determine which parts of the system do not bring value to the workflow (Slack & Brandon-Jones, 2021). Creating a continuous workflow means ensuring each team's workflow progresses smoothly and preventing any interruptions or bottlenecks that may occur with cross-functional teamwork. Kanban, a lean management technique that utilizes a visual cue to trigger action, is used to enable easy communication between teams so they can address what needs to be done and when it needs to be done. Breaking the total work process into a collection of smaller parts and visualizing the workflow in this regard facilitates the feasible removal of process interruptions and roadblocks. Developing a pull system ensures that the continuous work-flow remains stable and guarantees that the teams deliver work assignments faster and with less effort. A pull system is a specific lean technique that decreases waste in any production process. It ensures that new work is only started if there is a demand for it, thus minimizing overhead and optimizing storage costs. The last principle is continuous improvement and can be regarded as the most important step in the lean management method. Facilitating continuous improvement refers to a variety of techniques that are used to identify what an organization has done, what it needs to do, any possible obstacles that may arise, and how all members of the organization can make their work processes better. The lean management system is neither isolated nor unchanging; therefore, issues may occur within any of the other four steps. Ensuring that all employees contribute to the continuous improvement of the workflow protects the organization whenever problems emerge. The management has to create an environment and culture in which all employees can work in line with the five principles (Helmold *et al.*, 2022).

1.2 Lean Management: Principles of Kaizen, Kata, and Keiretsu

1.2.1 Definition and Scope

The lean management concept and philosophy are based on continuous improvements across the value chain (Kaizen). In addition, there are two additional elements that play a significant role in successful transformation. These two additional elements are Kata and Keiretsu (Figure 1.2). In Japanese, Kata (型) stands for a set of structured routine movements that a martial arts practitioner would deliberately engage in to ensure that the specific moves become second nature. Keiretsu refers to external partnerships with suppliers, stakeholders, and members in the value chain (e.g., auditors, laboratories, and distributors). A Keiretsu network

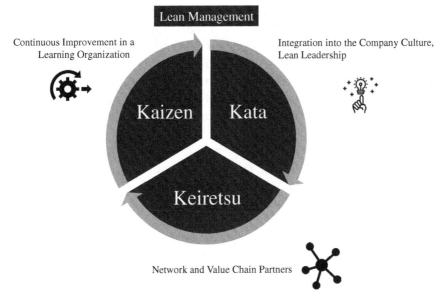

Figure 1.2. Kaizen, Kata, and Keiretsu.

(系列 ネットワーク), or Keiretsu value chain network (integration, order, or system of stakeholders, partners, and suppliers), represents a means of mutual security within the lean management methodology and usually includes large manufacturers and their suppliers of raw materials, systems, and components (Ahmadin & Lincoln, 2001).

1.2.2 Kaizen

Kaizen (改善) is a Japanese management concept that targets improvements in small steps. According to this concept, all personnel are expected to stop their work when they encounter any abnormality and, along with their supervisor, suggest an improvement to resolve the abnormality. In Kaizen, quality is the aim of daily life, not only during working hours. Improvement should be gradual and infinite. One should pursue perfection. Employees should be continuously engaged in the company's life and in the improvement of every aspect of the company (processes, products, infrastructure, etc.). This improvement throughout all aspects of life is related to the great attention that is paid to the needs and requirements of the customer.

Kaizen focuses on teams (quality circles) and promotes teamwork and team spirit; however, it also recognizes individual contribution. It emphasizes the engagement of each worker to the concept and vision of the company, so that employees will identify themselves with the enterprise, its culture, and objectives. The important aspects of Kaizen are as follows:

- Understand what is wrong, not who is wrong.
- How do we eliminate waste (Muda)?
- How do we decrease quality costs?

The term Kaizen is Japanese and can be translated to "change for the better." The main goal of Kaizen is to continuously improve areas of work, processes, and products by integrating the people of the affected areas. Usually, Kaizen is realized through workshops. The typical duration of a workshop can vary from three to five days. Kaizen must be introduced in daily routines and are supported through improvement-oriented workshops. The aim of Kaizen is to implement methods, tools and principles in the day-today work (Liker, 2020). The 10 principles of Kaizen can be described as follows:

- Say no to the *status quo*.
- If something is wrong, correct it.
- Accept no excuses, and make things happen.
- Improve everything continuously.
- Abolish old, traditional concepts.
- Be economical.
- Empower everyone to take part in problem solving.
- Before making decisions, ask "why" five times to get to the root cause.
- Get information and opinions from multiple people.
- Remember that improvement has no limit; never stop trying to improve.

A useful tool in the context of Kaizen is the P-D-C-A cycle. PDCA is an iterative four-step management method used in business for the control and continuous improvement of processes and products. It is also known as the Deming circle/cycle/wheel or the Shewhart cycle. The PDCA cycle is a method, which has been used since the 1950s. It is a tool, which has the following improvement steps:

- *Plan*: Analyze the current situation and define the improvement plan.
- *Do*: Implement the defined solutions.
- *Check*: Evaluate the improvement results.
- *Act*: Define the counter-actions in case of a deviation from the objective; standardize the best solution.

After improvement, it is important to standardize and implement the action, so that the process or activity does not return to the old state. If this is ensured, one can aim for the next improvement.

1.2.3 Kata

The term Kata comes from the Japanese martial arts and describes standardized and repeated movements. Through practice and application, this philosophy becomes routine and internalized to such an extent that it is reflexively carried out without thinking. Transferred to the management of organizations or companies, Kata means the development and anchoring of routines in thinking and acting on all organizational levels.

The aim of Kata coaching, as developed by the Japanese car manufacturer Toyota against this background, is to anchor and increase the competence for problem solving as a core ability in the entire company by developing appropriate learning and behavioral routines. In addition, regardless of people, companies must incorporate methods to continuously question processes and activities, improve them systematically and sustainably (continuous improvement process, CIP), and be able to react flexibly to new things (learning organization). Employees should develop the ability to deal productively and creatively with uncertainties, problems, and changes and thereby increase the company's performance.

1.2.4 Keiretsu

A Keiretsu network (系列 ネットワーク), or Keiretsu value chain network (integration, order, or system of stakeholders, partners, and suppliers), represents a means of mutual security, especially in Japan, and usually includes large manufacturers and their suppliers of raw materials, systems, and components (Ahmadin & Lincoln, 2001). Keiretsu groups are clusters of independently managed firms maintaining close and stable economic ties, cemented by a governance mechanism, such as presidents' clubs, partial cross-ownership, and interlocking directorates. Within this broad definition lie two distinctive variations. A horizontal Keiretsu network refers to conglomerates covering several industries linked by cross-shareholding, intra-group financing, and high-level management by a central (often inconspicuous) body of directors. A vertical Keiretsu network refers to groups surrounding one big manufacturer and consists of a multi-layered system of suppliers focused on the core company. Keiretsu networks have received much attention in the European automotive and transportation sectors due to the success of Japanese companies such as Toyota, Mitsubishi, and Hitachi and other conglomerates in achieving improved customer service, better inventory control, and more efficient overall channel management (Freitag, 2007). Keiretsu, which is a form of Japanese business network, shares many goals of several business functions. The concept of a Keiretsu network was introduced by Toyota in the mid-1980s (Helmold *et al.*, 2022) and transferred to affiliates and suppliers outside Japan.

Keiretsu networks often include partial ownership of the respective supplier as collaborative Keiretsu or supply (Helmold *et al.*, 2022). Control relationships between pairs of firms represent a form of bilateral exchange. The philosophy of Keiretsu may lead to broad functional and cultural changes for those companies which use the system (Helmold *et al.*, 2022). Keiretsu networks with financial and commercial connections develop quasi-administrative ties through cross-shareholding, as Keiretsu networks have two sides: (1) horizontal relationships based on mutual support and (2) vertical structures based on asymmetric exchange and control between financial firms and industrial firms. In various articles and books, Liker (2004) explains the Toyota Way and the principles of Keiretsu supply networks. Many OEMs and their suppliers have meanwhile adopted this system (Liker & Choi, 2005). The Keiretsu system has the following points:

- Keiretsu refers to the Japanese business structure comprising a network of different companies, including banks, manufacturers, distributors, and supply chain partners.
- Before the Keiretsu system, the primary form of corporate governance in Japan was zaibatsu, which referred to small, family-owned

businesses that eventually evolved into large, monopolistic holding companies.

- Horizontal Keiretsu refers to an alliance of cross-shareholding companies led by a Japanese bank that provides a range of financial services.
- Vertical Keiretsu is a partnership of manufacturers, suppliers, and distributors that work cooperatively to increase efficiency and reduce costs.
- A drawback of the Keiretsu system is the easy access to capital, which can lead a company to take on too much debt and invest in risky strategies.

1.3 Lean Management Versus Traditional Concepts

In contrast to traditional manufacturing concepts and value chain activities, lean management is based on a reduction in lead time, low inventory levels, and the permanent elimination of non-value-adding activities along the value chain (Ohno, 1990). These (non-value-adding) activities are unnecessary and represent waste, or "Muda" (無駄). Figure 1.3 shows the two concepts, the traditional and the lean. Both concepts are aimed at customers. The focus of the lean management concept is on optimal

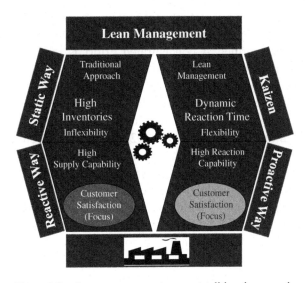

Figure 1.3.   Lean management versus traditional approach.

responsiveness and not on inventory. Inventories increase the cost of capital and negatively impact shareholder value, while short cycle times result in low inventories. Lean manufacturing or lean production, often simply called "lean," is a systematic method of eliminating waste ("Muda") within a manufacturing system. The lean management concept also takes into account the fact that waste can arise from overloading ("Muri") or from an imbalance in workload ("Mura"). From the perspective of the customer consuming a product or service, "value" is any action or process that a customer would be willing to pay for. Essentially, lean is about making visible what creates value by reducing everything else. Lean manufacturing is a management philosophy derived primarily from the TPS (the term Toyotaism is also widely used), which was only identified as "lean" in the 1990s. TPS is known for its focus on reducing the original seven Toyota wastes to improve overall customer value; however, there are different perspectives on how best to achieve this. Toyota's steady growth from a small company to the world's largest automaker has drawn attention to how it achieved this success (Helmold *et al.*, 2022).

1.4 Kaizen and Kaikaku

Kaizen is continuous improvement in small steps. But, in reality, improvement is not uniform, but discontinuous, sometimes faster and sometimes slower. With Kaizen, we sometimes face major obstacles and get stuck. In this case, we have to change the system, the framework conditions, in order to improve further. A concentrated effort is required to make a breakthrough in Kaizen. This method is referred to as Kaikaku and helps us to improve quality in production (改革, reformation, or transformation). Kaikaku is called "breakthrough Kaizen" or "Kaizen lightning." Kaikaku means "revolution in thought and action" and "improvement of great significance" (Helmold *et al.*, 2022). With a concentrated use of force, it temporarily accelerates the speed and scope of Kaizen. Kaizen is the continuous improvement of an activity with the purpose of increasing the performance in the production system, usually less than 20% or 30%, in a given period of time (Helmold *et al.*, 2022). Kaikaku is the opposite of Kaizen as it represents drastic change or radical improvement, usually resulting from big investments in technology and/or equipment. This way of constantly making improvements to processes may seem impossible. So, there is a need to initiate a radical change in the company (Helmold *et al.*, 2022). Kaikaku is not like Kaizen, which is engaged in by all employees; Kaikaku usually starts with the top management of the company and then moves to the lower levels of worker (Yamamoto, 2013). This is because Kaikaku represents crucial strategic changes. Table 1.1

Table 1.1.   Kaizen versus innovation.

Kaizen (改善)	Innovation
Small steps	Major change
Low investment	High investment
No risk	Entrepreneurial risk
Involvement of people	Specialized team
Short term	Long term
Team decisions	Management decision

Table 1.2.   Differences between Kaizen and Kaikaku.

Kaizen (改善)	Kaikaku (改革)
Lots of small, incremental improvements in all areas of production and other value chain processes.	Major and selective improvements aimed at more complex and strategic projects.
Systematic analysis of all areas and projects in the value chain.	Systematic analysis and synthesis of individual and selected projects and improvement measures.
Involvement of many teams and individuals across all functions.	Involves fewer people, as only selective projects are implemented.
Starts with management and involves all employees as part of the Kaizen culture.	The participating organizational parts of selected projects are geared toward the improvement goals.
Uses and applies mostly tactical and operational measures.	Driven by top management with strategic measures.
Continuously proceeding in all areas with improvements.	Limited to a certain time and duration for selected projects.

compares Kaizen and innovation (Helmold *et al.*, 2022). Table 1.2 shows the differences between Kaizen and Kaikaku.

References

Ahmadin, C. & Lincoln, E. J. (2001). Keiretsu, governance, and learning: Case studies in change from the Japanese automotive industry. *Organization Science*, 12(6), 683–701.

Belekoukias, I., Garza-Reyes, J. A., & Kumar, V. (2014). The impact of lean methods and tools on the operational performance of manufacturing organisations. *International Journal of Production Research*, 52(18), 5346–5366.

Bertagnolli, F. (2020). *Lean Management. 2. Auflage. Einführung und Vertiefung in die japanische Man-agement-Philosophie*. Wiesbaden: Springer.

Helmold, M. & Terry, B. (2021). *Operations and Supply Management 4.0 Industry Insights, Case Studies and Best Practices*. Cham: Springer.

Helmold, M. *et al.* (2022). *Lean Management, Kaizen, Kata and Keiretsu. Best-Practice Examples and Industry Insights from Japanese Concepts*. Cham: Springer.

Liker, J. K. (2004). *The Toyota Way*. Madison: Mc Graw-Hill.

Ohno, T. (1990). *Toyota Production System. Beyond Large Scale Production*. New York: Produc-tivity Press.

Slack, N. & Brandon-Jones, B. A. (2021). *Slack: Operations and Process Management. Principles and Practice for Strategic Impact* (6th edn.). London: Pearson.

Yamamoto, J. J. (2013). *Kaikaku in Production Toward Creating Unique Production Systems*. Västerås: Mälardalen University.

Chapter 2

Lean Management in Japan

2.1 Toyota Production System (TPS) Conquering the World

As Kiichiro Toyoda, Taiichi Ohno, and others at Toyota reflected on their production system in the 1930s, and more intensely just after World War II, it occurred to them that a series of simple innovations might make it more possible to provide both continuity in process flow and a wide variety in product offerings (Ohno, 1990). They therefore revisited Ford's original thinking, and invented the Toyota Production System (TPS, Toyota Seisan Hōshiki — トヨタ生産方式). This system in essence shifted the focus of the manufacturing engineer from individual machines and their utilization to the flow of the product through the total process (Furata, 2021). Toyota concluded that by right-sizing machines for the actual volume needed, introducing self-monitoring machines to ensure quality, lining the machines up in process sequence, pioneering quick setups so that each machine could make small volumes of many part numbers, and having each process step notify the previous step of its current needs for materials, it would be possible to obtain low cost, good variety, high quality, and very rapid throughput times to respond to changing customer desires. The concept of the TPS is based on a paradigm of permanent and continuous improvement, the Kaizen philosophy (Pascual, 2013). Figure 2.1 displays the meaning of Kaizen as "change for the better or change for improvement." Also, information management could be made much simpler and more accurate (Liker, 2004). Kaizen must be incorporated as part of the mission and vision of the enterprise. Together with

Figure 2.1.   Kaizen — Change for the better.

Kata (methods, or routines) and Keiretsu (networks), lean management is the ideal philosophy and paradigm to focus on value-adding activities across the value chain and with all relevant stakeholders (Helmold *et al.*, 2022).

The thought process of lean was thoroughly described in the book *The Machine that Changed the World* (1990) by Womack, Jones, and Ross. The authors described that lean principles are based on the following five elements:

(1) Specify the value desired by the customer.
(2) Identify the value stream for each product providing that value and challenge all of the wasted steps (generally 9 out of 10) currently necessary to provide it.
(3) Ensure a continuous flow of the product through the remaining value-added steps.
(4) Introduce pull between all steps where continuous flow is possible.
(5) Strive toward perfection so that the number of steps and the amount of time and information needed to serve the customer continually falls.

2.2 Lean Management Evolution

The evolution of lean management can be seen in Figure 2.2. This continued success has over the past two decades created an enormous demand for greater knowledge about lean thinking. There are literally hundreds of books and papers, not to mention thousands of media articles exploring the subject, and numerous other resources available to its growing audience. As lean thinking continues to spread to every country in the world, leaders are also adapting the tools and principles beyond manufacturing, to logistics and distribution, services, retail, healthcare, construction, maintenance, and even governance (Bertagnolli, 2020). Indeed, lean consciousness and methods are only beginning to take root among senior managers and leaders in all sectors today (Belekoukias *et al.*, 2014). Value chain networks in the present are complex and international structures of supply and demand. Especially, Japanese makers show how suppliers are sustainably integrated into their own value chain and activities (Helmold & Terry, 2021). The Japanese networks are described as "Keiretsu networks," in which suppliers and customers are integrated systems throughout the value chain (Helmold & Samara, 2019). Future lean management concepts and supply chains will be configured in a transparent and optimal way, so

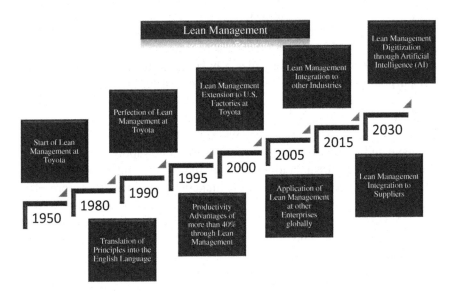

Figure 2.2.   The evolution of lean management.

that wasteful activities and processes can be eliminated at the earliest point of time (Srai & Gregory, 2008). In the future, competitiveness will be decided by who has the most flexible and efficient value network, including value streams from raw material suppliers over a company's own operations to the distribution to the customers (Helmold & Terry, 2021).

2.3 Lean Thinking as Part of the Japanese Society

Lean management is part of Japanese society. It involves the concept of Kata rules and guidelines. These are general societal rules and patterns of behavior that the Japanese exhibit in their everyday life. Since this is an ingrained part of Japanese culture, the standard practices of business come very naturally to people in this country. As part of their culture, the Japanese are also known for being perfectionists in everything that they do. They take a great deal of pride in their work, so they take a great deal of care in training their employees to ensure the highest quality of work. Another big part of this concept of lean is listening and patience. Toyota is a company that has been heavily associated with lean. One of the things that makes this company so unique is that rather than just barking orders, they encourage workers to develop their own skills and problem-solving abilities. There is also a push to focus more on the long-term goals of the company rather than the short-term. These are all areas that exemplify how crucial patience is to Japanese culture. In fact, to be considered an expert in any field it takes a great deal of time, so patience is necessary. Lean is all about how you organize a business to make it more efficient. The purpose is to maximize value and reduce waste. The Japanese culture is known for valuing cleanliness and order, both in the way they live their everyday life and in how they run their businesses. If everything is meticulously organized, you can reduce waste because you know exactly what you have and what you need. This applies to office supplies and workflow management. You are able to see who should be working on what in order to know what needs to get done. It makes the business run more efficiently. Figure 2.3 shows how lean tools of visualization are integrated into Japanese society and life. It shows the Tozai line, including information on connections, time, and also the location. Other lines and connections are marked in different colors. Another example of lean management and artificial intelligence is a bakery in Tokyo (Figure 2.4), where the customer-selected products are identified through a camera on a special

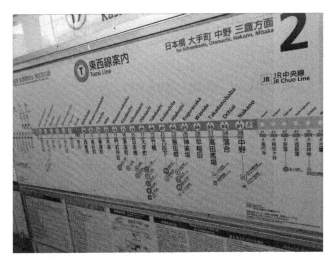

Figure 2.3. Visualization and lean processes (Tozai line).

Figure 2.4. Lean processes in a bakery.

counter. The staff only needs to confirm, so that the price is shown to the customer. The customer can now pay via telephone payment (one scan), via card, or via inserting money into a slot. The change is given automatically. The process is very fast; wasteful elements such as waiting time are eliminated, and the staff can use more time to advise and assist customers.

2.4 The Concept of Monotsukuri

The Japanese concept of Monotsukuri (ものづくり) is a combination of "mono," meaning thing, and "zukuri," which is the act of making. It simply means craftsmanship or manufacturing and is used as a buzzword in industry and mass media to embody the Japanese spirit and history of manufacturing. It is a word of Japanese origin and has only recently, since the latter half of the 1990s, come to mean manufacturing and production. The broader meaning of *monozukuri* encompasses a synthesis of the technological prowess, know-how, and spirit of Japan's manufacturing practices. The spirit includes a sincere attitude toward production with pride, skill, and dedication, and the pursuit of innovation and perfection. This concept is definitely not a burden or something pushed by a supervisor to the front line. It is a way of pursuing innovation and perfection, often disregarding profit or the balance sheet.

This mindset of perfection has been integrated in Japanese culture for centuries. Many elements are much older than the Toyota Production System. Examples of this mindset are visible in Japan in many situation of daily life. This perfection was initiated much earlier than the TPS literature.

2.5 Impacts of Bushido on Lean Management

2.5.1 Scope and Definition

Bushido (Japanese: 武士道) signifies the code of ethics and ideals that dictated the *samurai* way of life in ancient Japan. The moral values of *samurai* warriors stress elements such as sincerity, frugality, loyalty, martial arts, and honor until death. Bushido flourished during the Edo period from 1600 to 1878. Inspired by Neo-Confucianism during the Edo period and influenced by Shinto and Buddhism, Bushido allowed the *samurai* to be tempered by wisdom, patience, and serenity. The seven virtues are shown in Figure 2.5.

2.5.1.1 *Justice or Rectitude (Gi)*

Gi is all about making sure that we have the right way when we make a decision and the power to make a decision quickly. It is about making sure

Bushidō (武士道)

Figure 2.5.   Bushido and lean management.

that we do not become indecisive and that our decisions are made based on the right reasons.

2.5.1.2 *Courage* (*Yū*)

Yū is about making sure what we do is right and that we have the courage to do the right thing, not just what people think we should do. If we are raised in a particular way, we think in a way that we believe in. This is about making sure we do what we believe in and have the courage to do so.

2.5.1.3 *Compassion or Mercy* (*Jin*)

As a warrior, the *samurai* had the power to kill. However, benevolence is about making sure that you are balanced in how you think. It is about making sure that you also have sympathy and mercy at the right time. For the *samurai*, it was about making sure he fought for the right reason and that if he had to kill someone, he not only did it for the right reason and his belief but also made sure that if there was no need to kill, he would have shown mercy and sympathy.

2.5.1.4 *Respect* (*Rei*)

According to Rei, it is important to show respect for everything we believe and be polite to everyone. The way the *samurais* lived their lives meant they must be respectful to their elders, respect life, and respect others' beliefs.

2.5.1.5 *Honesty* (*Makoto*)

Honesty was very important to the *samurais*, as they believed that being honest in everything you do gives you respect and means you can be trusted.

2.5.1.6 *Honor* (*Meiyo*)

To live and die with honor was very important to a *samurai*, and they did everything they believed in with honor.

2.5.1.7 *Loyalty* (*Chūgi*)

Loyalty was probably one of their most important attributes. They treated each other like family and would do everything within their power to protect and help other *samurai* warriors. Loyalty was important because this meant they could trust other warriors and know they would be loyal to whatever they needed to do and not worry about losing their respect. Loyalty was even shown after the death of a person. Figure 2.7 shows a funeral meeting (Jap.: Sogi Houjiki 葬儀ほうじき). Nearly all Japanese funerals are conducted in Buddhist-style, regardless of what religion the family practices. Also, almost all involve cremation, to the point that even the local government sometimes bans traditional burials. The Ososhiki is the actual Japanese funeral service and contains several ceremonies. It starts one day after the Otsuya with a Sougi or Soshiki, which is the funeral ceremony itself. It follows a similar procedure to the Otsuya, with a priest chanting a *sutra* and the bereaved burning incense. After the Sougi, there is a Kokubetsushiki, or memorial ceremony, where the friends and acquaintances of the bereaved pay their respects to the dead and offer condolences to the family. Lastly, there is a cremation ceremony. This is a very private ceremony conducted only within the family. After cremation, the family uses chopsticks to pick the bones out of the ash and place them in a burial urn, which is then interred inside the family grave. Black is the color of mourning in Japan. Lately, dark blue and dark gray are becoming more acceptable, but black is still the most preferred color. It can't be stressed enough that you must never attend a Japanese funeral without the appropriate attire — to do so would be the ultimate sign of disrespect.

2.6 Ikigai as Part of the Lean Management Philosophy

Ikigai (Jap.: 生 き 甲 斐, meaning of life) (see Figure 2.6) is a Japanese concept that means the reason for being. "Iki" in Japanese means life, and "gai" describes value or worth. The *ikigai* is the life purpose or the bliss of a person or group. It's what brings you joy and inspires you to get out

Figure 2.6. Ikigai.

Figure 2.7.   Death funeral.

of bed every day. Ikigai is loosely translated as "that which is worth living for," "joy and the goal in life," or, to put it casually, "the feeling of having something for which it's worth getting up in the morning." In Japanese culture, the often lengthy and thorough self-exploration in the pursuit and search for *ikigai* has an important meaning. It is a very personal process, and the result can therefore vary from person to person. If a person finds or has his *ikigai*, it gives him a feeling of *joie de vivre*, i.e., inner satisfaction. In connection with the self-understanding of the cultural identity of Japanese society, discussions are held in the media on which social ideals should serve as the basis for *ikigai*, what can be regarded as *ikigai* (and what not), and Ikigai is the way consisting of two aspects. The first element is sources or objects that bring value or meaning to life. The second element is a feeling that one's life has value or meaning because of the existence of its sources or objects (Inoue, 2000). Figure 2.6 depicts the concept. There are four dimensions to find *ikigai*: (1) what one loves doing, (2) what the world needs, (3) what one is paid for, and (4) what one is good at. The combination of these four categories is considered "Ikigai" and these lead to the mission, the vocation, the profession, and the passion of individuals and groups. Figure 2.7 shows a death funeral in Japan, in which dead family members and friends are valued.

References

Belekoukias, I., Garza-Reyes, J. A., & Kumar, V. (2014). The impact of lean methods and tools on the operational performance of manufacturing organisations. *International Journal of Production Research*, 52(18), 5346–5366.

Bertagnolli, F. (2020). *Lean Management. 2. Auflage. Einführung und Vertiefung in die japanische Management-Philosophie*. Wiesbaden: Springer.

Furata, K. N. (2021). *Welcome Problems, Find Success Creating Toyota Cultures Around the World*. New York: Productivity Press.

Helmold, M. & Samara, W. (2019). *Progress in Performance Management. Industry Insights and Case Studies on Principles, Application Tools, and Practice*. Cham: Springer.

Helmold, M. & Terry, B. (2021). *Operations and Supply Management 4.0 Industry Insights, Case Studies and Best Practices*. Cham: Springer.

Helmold, M. *et al.* (2022). *Lean Management, Kaizen, Kata and Keiretsu. Best-Practice Examples and Industry Insights from Japanese Concepts*. Cham: Springer.

Inoue, K. (2000). *Psychology of Aging. Chuo Hoki Shuppan*. pp. 80–99, 144–145.

Liker, J. K. (2004). *The Toyota Way*. Madison: McGraw-Hill.

Ohno, T. (1990). *Toyota Production System. Beyond Large Scale Production*. New York: Productivity Press.

Pascual, M. D. (2013). Toyota: Understanding the Key to Success: Principles and Strengths of a Business Model. Quezon City: Pluma Publishing.

Chapter 3

Lean Policy Deployment — Strategic and Cultural Transformation

3.1 Transformation to a Lean Enterprise — Hoshin Kanri

3.1.1 Scope and Definition

Lean policy deployment (Hoshin Kanri (方針管理)) is a method and tool (Figure 3.1; Work Aid 03) for ensuring that a company's strategic lean goals drive progress and action at every level within the company's value chain (Helmold et al., 2022). This method eliminates the waste that comes from inconsistent direction and poor communication. Hoshin Kanri strives to get every employee pulling in the same direction at the same time. It achieves this by aligning the goals of the company (strategy) with the plans of middle management (tactics) and the work performed by all employees (operations). Hoshin Kanri starts with a strategic plan (e.g., an annual plan) that is developed by top management to further the long-term goals of the company (Bertagnolli, 2020).

Hoshin Kanri is a step process used in strategic planning in which strategic goals are communicated throughout the company and then put into action. The Hoshin Kanri strategic planning system originated in post-war Japan but has since spread to the U.S. and around the world. Translated from Japanese, Hoshin Kanri aptly means Compass Management. The individual words "hoshin" and "kanri" mean direction and administration, respectively. Hoshin Kanri requires a strategic mission and vision in order to succeed.

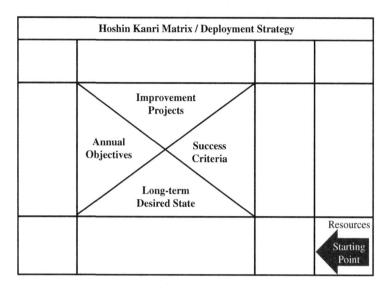

Figure 3.1.   Hoshin Kanri matrix.

From there, strategic objectives need to be clearly defined, with goals being written for long periods of a one- to five-year-long timeframe. Once the long-term timeframe goals are completed, the team can focus on yearly objectives. Management needs to avoid picking too many vital goals in order to stay focused on what is strategically important. The big goals then need to be broken down into smaller goals at a weekly and monthly basis and then implemented so that everyone, from management to the factory floor, is in agreement on what needs to be accomplished. The satisfaction of goals should be reviewed on a monthly basis, with a larger annual review at the end of the year. Performance measurement is also a key part of the process. Hoshin Kanri is a top-down approach, with the goals being mandated by management and the implementation being performed by employees. As a result, systems need to be in place to ensure that objectives from senior management are effectively communicated all the way down the hierarchy chain of the enterprise (Ahmadin & Lincoln, 2001). As such, a catchball system is often used in order to aid in the execution of the strategic plan. A catchball system seeks to get the opinions of both managers and employees through meetings and interactions in order to ensure the bidirectional flow of goals, feedback, and other information throughout the organization. There are four quadrants found within this planning tool as shown in Figure 3.1:

(1) Long-term desired state
(2) Annual and short-term objectives
(3) Improvement projects
(4) Success criteria

3.1.2 Process

The process you use to develop your strategic plan is as important as the plan itself. That's why many organizations turn to the Hoshin Kanri approach, which focuses on creating a plan that takes into account both the daily management of the organization and the tactics necessary to reach those goals that will have the most significant impact. The result is a set of specific action plans and resources necessary to achieve your business breakthrough.

Although some organizations tweak the approach to meet their specific needs, most often, the Hoshin Kanri (or Hoshin planning) process consists of the following seven steps (Table 3.1):

Step 1: Establishing the Lean Mission, Vision, and Assessing the Current State

During this phase, it is important to examine the targeted lean mission, vision, and values. Are they aligned with a realistic future state? It is also necessary to review existing processes and procedures that are designed to meet future objectives.

Step 2: Developing Breakthrough Objectives

Breakthrough objectives are those that require the organization to stretch in new and challenging ways. They typically take three to five years

Table 3.1.   Hoshin Kanri process.

No.	The seven steps of Hoshin Kanri planning
1.	Establishing a lean mission and vision
2.	Developing strategic lean objectives
3.	Developing annual lean objectives
4.	Cascading goals throughout the organization
5.	Executing annual objectives
6.	Implementing monthly lean reviews
7.	Conducting annual reviews

to achieve. Breakthrough objectives often include entering new markets, introducing new products, or adopting a different service delivery model.

Step 3: Defining Annual Objectives
What needs to happen this year to ensure the next three to five year goals are met? For example, if the goal is to introduce a new product in three years, it may be necessary to complete a market research study and define the product requirements during this year.

Step 4: Cascading Goals Throughout the Organization
Once you know what you need to accomplish, you can begin to assign department, team, and individual objectives that align with the overall mission for the year. The goals should be measurable and specific, with defined key performance indicators that can be monitored by leadership.

Step 5: Executing Annual Objectives
In this step, action is taken to reach the goals for the year. A host of problem-solving techniques, including Kaizen events, DMAIC, PDSA, and A3, can be applied to help ensure success.

Step 6: Conducting Monthly Reviews
Too often, organizations spend a bunch of time setting up a plan for the year, only to forget about it as the pressure of daily management mounts. The Hoshin Kanri process helps avoid this common problem with monthly reviews. Checking in on each person's progress toward goals on a monthly basis ensures that forward progress is maintained.

Step 7: Doing Annual Review
At the conclusion of 12 months, it is time to do a comprehensive assessment of the organization's progress. It may be necessary to make adjustments to goals or time estimates. This is also a good time to ensure that resources are properly allocated for what needs to be accomplished in the next year.

Hoshin Kanri (or Hoshin planning) is not terribly complex, but each step is important to create an action plan that can be effectively executed in the expected timeframe. Aligned goals ensure that everyone is working toward the same ends, and frequent reviews help ensure that work stays on track. If you are looking for a planning method that will help your organization make the leap to the next level, Hoshin Kanri is a great option to consider.

3.2 Strategic Triangle

3.2.1 Definition and Scope

Strategic lean management is the process of successfully transforming into a lean company to achieve a competitive advantage. This transformation starts with three questions, as shown in Figure 3.2:

- Where are we in terms of lean management?
- Where do we want to go in terms of lean management?
- How do we successfully achieve lean transformation?

The process of strategic lean management is a sequential process with three elements, as outlined in the Strategic Lean Management Triangle in Figure 3.3 (Helmold *et al.*, 2022). The three steps comprise the (1) lean analysis, (2) lean options, and (3) lean implementation.

3.2.2 Lean Analysis

The lean analysis of an organization is about understanding the strategic position of the organization in terms of lean management. This stage requires a profound analysis of where the organization stands in terms of

Figure 3.2.   Strategic questions.

Lean Management Strategic Triangle

Figure 3.3.   Strategic triangle.

lean management tools and processes (Johnson *et al.*, 2017). The existing competencies and resources of the organization need to be assessed to determine if there are any opportunities to be gained from them and if they need to be enhanced in order to pursue strategic objectives and goals. The opinions or viewpoints of major stakeholders who influence the organization must be taken into account, as the purpose of all of the strategic analysis is to define the potential future direction of the organization. The purpose of this phase (strategic analysis) is to create a suitable starting position and to understand the key influences on the present and future state of the organization, what opportunities are afforded by the environment, and the competencies of the organization.

Since strategy is concerned with the position a business takes in relation to its environment, an understanding of the environment's effects on an organization is of central importance. The historical and environmental effects on the business must be considered, as well as the present effects and the expected changes in environmental variables. The analysis of the environment can be done via macro- and microanalysis. Additionally, strengths, weaknesses, opportunities, and threats (SWOT) complete the assessment of the environment. This step is a major task because the range of environmental variables is so vast. Another area of strategic analysis is the evaluation of the strategic capability of an organization and where it is able to achieve a competitive advantage. Considering the resource areas of a business, such as its physical plant, its management, its financial structure, and its products, one may identify these strengths and weaknesses. The expectations of stakeholders are important because they will affect what will be seen as acceptable in terms of the strategies advanced by management. Stakeholders can be defined as people or groups inside or outside the organization, who have an interest in the activities of the organization. A typical list of stakeholders for a large company would include shareholders, banks, employees, managers, customers, suppliers, governments, and society. Culture affects the interpretation of environmental and resource influences (Helmold & Terry, 2021).

3.2.3 Strengths and Weaknesses

The SWOT analysis is a framework used to evaluate a company's competitive position and develop strategic planning. SWOT analysis assesses internal and external factors, as well as current and future potential. This technique, which operates by peeling back layers of the company, is designed for use in the preliminary stages of decision-making processes and can be used as a tool for evaluating the strategic position of organizations of many kinds (for-profit enterprises, local and national governments, NGOs, etc.). It is intended to specify the objectives of the business venture or project and identify the internal and external factors that are favorable and unfavorable to achieving those objectives. Users of a SWOT analysis often ask and answer questions to generate meaningful information for each category to make the tool useful and identify their competitive advantage.

3.2.4 Analyzing the Core Competencies

The core competency concept describes a product, feature, process, skill, brand, or activity that a company can perform better than the competition

Figure 3.4.   Strategic mission and vision.

and has thus achieved a competitive advantage. It is determined by the certain characteristics like customer value or benefits, protection against imitation, differentiation, diversification, and innovation or unique features, as shown in Figure 3.4. In business, a competitive advantage is the attribute that allows an organization to outperform its competitors.

3.2.5 Lean Options and Strategic Choice

Lean options and the lean strategic choice typically follow strategic analysis. Strategic choice involves a whole process through which a decision to choose a particular option from various lean alternatives in the value chain and roll-out plan is taken. There can be various methods through which the final choice can be selected. Managers and decision-makers keep both the external and internal environment in mind before narrowing it down to one. It is based on the following three elements: first, the generation of strategic options, e.g., growth, acquisition, diversification, or concentration; second, the evaluation of the options to assess their relative merits and feasibility; third, the selection of the strategy or option that the organization will pursue. There could be more than one strategy chosen, but there is a chance of an inherent danger or disadvantage to any choice made. Although there are techniques for evaluating specific options, the selection is often subjective and likely to be influenced by the values of managers and other groups with an interest in the organization (Helmold *et al.*, 2022).

The generic strategies of lean differentiation and lean cost leadership are a good method to define in which direction a company should go to increase profitability and to acquire a competitive advantage.

3.2.5.1 *Horizontal lean strategy diversification*

Horizontal diversification describes the development of a new product that is still factually related to the product range previously offered. The existing value chain can simultaneously be used with minimal adjustments. With horizontal diversification, a company expands its offerings at the same economic level to reach new customers. An example of this type of diversification is the development of the iPad, which with its introduction gradually expanded Apple's existing smartphone and computer portfolio.

3.2.5.2 *Vertical lean strategy diversification*

With vertical diversification, a company deepens its commitment to sales-oriented activities (forward integration) and/or the actual manufacturing process of its products (backward integration). Diversification does not take place at the same level of the value chain as with horizontal diversification but upstream or downstream. With forward integration, a company takes the sales of its products and services into its own hands, for example, by opening its own branches or an online shop. Backward integration describes the safeguarding of a company's reference markets, for example, by taking over production processes that were previously outsourced to external companies. While horizontal diversification aims to reduce dependency on one product line, vertical diversification focuses on reducing dependence on suppliers and dealers. The acquisition of the necessary skills and know-how for the successful implementation of sales and production processes is in turn associated with high investment costs and thus increased financial risks.

3.2.5.3 *Lateral lean strategy diversification*

Lateral diversification strategy helps the companies expand to completely new markets that have no material connection with their existing businesses.

The aim and purpose of this alignment are to minimize the dependence on developments in the existing market segment. Google can be mentioned as a good example in this context: In addition to the search engine core business, the company expanded early on into other market segments such as telecommunications (Fiber), biotechnology (Calico), or autonomous automotive technology (Waymo). The lateral diversification strategy is used by multinational companies, particularly to respond flexibly to changes and trends in the market. The necessary know-how is usually acquired through the acquisition of specialized companies that are already represented in the market of interest. Accordingly, this strategy requires enormous investment costs and harbors not only financial but also immaterial risks, such as a diluted brand image due to product offerings that are too diversified.

3.2.6 Lean Implementation

Strategic implementation is concerned with the transition of the selected strategy into action (Johnson & Scholes, 1997). The ways in which strategies are implemented are described as the strategic architecture or framework of the organization (Johnson & Scholes, 1997). Successful implementation of the chosen strategy will depend on several factors, such as stakeholder's expectations, the employees, the company culture, the will to change, and the cooperation within the organization (Belekoukias *et al.*, 2014). These elements and how the management and employees work together to adopt the new plan will decide how successful the strategy implementation is. The available skills and/or the ability to develop new skills when required for the planned change and issues like the structural reorganization and resulting cultural disturbance would also affect success. Resource availability and planning for the utilization of such resources need to be addressed as part of the implementation plan. The entire process necessitates the management of strategic change and concerns handling both hard and soft factors of the organization, i.e., structure and systems, culture and motivation, etc. Johnson and Scholes argue that for a strategy to be successful, it must satisfy three criteria (Johnson & Scholes, 1997). These criteria can be applied to any strategy decision, such as competitive strategies, growth strategies, or development strategies:

- Suitability: Whether the options are adequate responses to the firm's assessment of its strategic position.
- Acceptability: Considers whether the options are consistent with the firm's objectives and are acceptable to the stakeholders.
- Feasibility: Assesses whether the organization has the resources it needs to carry out the strategy.

3.2.7 Suitability

Suitability is a useful criterion for screening strategies. The following are questions related to strategic options:

Does the strategy exploit the company strengths, such as providing work for skilled craftsmen or environmental opportunities, e.g., helping to establish the organization in new growth sectors of the market?

How far does the strategy overcome the difficulties identified in the analysis? For example, is the strategy likely to improve the organization's competitive position, solve the company's liquidity problems, or decrease dependence on a particular supplier?

Does the option fit in with the organization's purposes? For example, would the strategy achieve profit targets or growth expectations, or would it retain control for an owner-manager?

3.2.8 Acceptability

Acceptability is essentially about assessing risk and return and is strongly related to the expectations of stakeholders. The issue of "acceptable to whom?" thus requires the analysis to be thought through carefully. Some of the questions that will help identify the likely consequences of any strategy are as follows:

How will the strategy impact shareholder wealth? Assessing this could involve calculations relating to profitability, e.g., net present value (NPV).

How will the organization perform in profitability terms? The parallel in the public sector would be cost/benefit assessment.

How will the financial risk (e.g., liquidity) change?

What effect will it have on the capital structure (gearing or share ownership)?

Will the function of any department, group, or individual change significantly?

Will the organization's relationship with outside stakeholders, e.g., suppliers, government, unions, and customers, need to change?

Will the strategy be acceptable in the organization's environment, e.g., higher levels of noise?

3.2.9 Feasibility

Feasibility assesses whether the organization has the resources it needs to carry out the strategy. Factors that should be considered can be summarized under the M-word model:

Machinery: What demands will the strategy make on production? Do we have sufficient spare capacity? Do we need new production systems to give lower cost/better quality/more flexibility, etc.?

Management: Is existing management sufficiently skilled to carry out the strategy?

Money: How much finance is needed and when? Can we raise this? Is the cash flow feasible?

Manpower: What demands will the strategy make on human resources? How many employees are needed, what skills will they need, and when do we need them? Do we already have the right people, or is there a gap? Can the gap be filled by recruitment, retraining, etc.?

Markets: Is our existing brand name strong enough for the strategy to work? Will new brand names have to be established? What market share is needed for success, and how quickly can this be achieved?

Materials: What demands will the strategy make on our relationships with suppliers? Are changes in quality needed?

Make-up: Is the existing organizational structure adequate or does it need to be changed?

3.3 Strategic Pyramid

3.3.1 Scope and Definition

A useful tool for transforming the enterprise successfully into a lean organization is the Strategic Pyramid in Figure 3.5 (Helmold *et al.*, 2022). Strategy in this context is the long-term lean management direction and positioning as well as the decision of the enterprise as to which business fields of the value chain to select and which lean strategies to choose. Strategy is therefore "the fundamental, long-term direction of three to five years and organization of a company in order to gain competitive advantages in a changing environment through the use of resources and competences and to realize the long-term goals of the stakeholders" (Johnson *et al.*, 2017).

3.3.2 Mission and Vision

Enterprises must manifest in their strategy the striving for lean excellence across the value chain (Helmold *et al.*, 2022). The mission is the starting point of the strategic pyramid. The mission (statement) of an enterprise is the long-term purpose of the company and the strategic lean directions

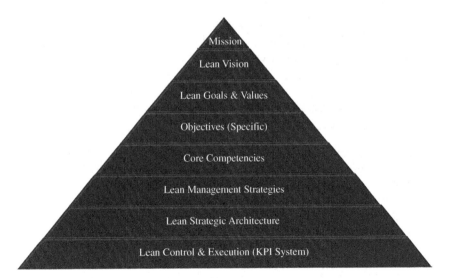

Figure 3.5.   Strategic pyramid.

across the value chain and value chain partners. The vision or strategic intent describes more specifically what an organization aims to achieve and its long-term aspirations (Johnson *et al.*, 2017).

3.3.3 Goals and Objectives

The mission and vision are followed by generic lean goals and specific lean objectives. Generic goals are not quantified and more general, but specific objectives are quantified and specific. Lean management objectives should be formulated in a specific, measurable, achievable, realistic, and timely manner.

3.3.4 Core Competencies

The next level in the strategic pyramid is the identification of core competencies. Core competencies are those competencies which allow companies to gain a superior or competitive advantage and that are very difficult for competitors to emulate (Johnson *et al.*, 2017). These describe the resources, skills, knowledge, or any other features that lead to a competitive advantage. Core competencies must be perceived by customers and clients (Helmold *et al.*, 2022). Lean knowledge should be acquired by establishing a lean culture and qualification plan. Lean excellence enterprises create their own lean academy and integrate value chain stakeholders into the lean culture and activities.

3.3.5 Lean Strategies

After defining the lean mission, vision, goals, objectives, and core competencies, the elements must be translated into strategic lean objectives and key performance indicators (KPIs). The long-term implementation of these elements is defined as the formulation of strategic objectives and is important for negotiations. In implementing the strategic goals, negotiations will take place with many stakeholders (Helmold *et al.*, 2022).

3.3.6 Lean Strategic Architecture

In addition to buildings, machines, plants, offices, resources, or employees, the infrastructure in the sense of lean management also includes knowledge and innovations of the company that ensure long-term success (Helmold *et al.*, 2022). This requires facilities, buildings, factories, or offices that represent the strategic infrastructure. In addition, however, other success criteria such as resources, knowledge, experts, name recognition, network, or innovations are of central importance. Companies can create a lean excellence environment through several academies:

- Lean Management Mission and Vision
- Lean Roll-out Plan across all Departments
- Lean Qualification Plan
- Lean Management Workshops
- Lean Management Academy
- Lean Management Experts
- Lean Management Audits and Workshops
- Lean Management Showrooms
- Lean Management Best Practices

3.3.7 Lean Control and Execution

The final element of the strategic pyramid is performance control (control and execution) and a target–performance comparison. A suitable tool for this step is the Balance Score Card (BSC) or an action plan. The instrument of the BSC was already developed in 1992 by professors Norton and Kaplan. The BSC is an instrument in strategic management and includes four categories:

- Customer Satisfaction
- Financial Category
- Internal Processes and Improvements
- Learning Organization

In practice, it seems that companies are adapting or expanding the original four dimensions to their specific needs (Johnson *et al.*, 2017). Figure 3.4 shows a practical example of Bombardier in China for mission and vision.

3.4 Lean Strategies Must Focus on Value-Creation

Porter postulated three generic or broad alternative strategies which may be pursued as a response to competitive pressures. They are termed generic strategies because they are broadly applicable to any industry or business. They are lean differentiation, cost leadership, and lean focus (Liker, 2004). A focus strategy may be further defined as cost focus, differentiation focus, or cost and differentiation focus. A differentiation strategy may be based on special lean processes, actual unique product features, or the perception thereof, conveyed through the use of advertising and marketing tactics, in the eyes of the customers (Ohno, 1990; Porter, 1985). Obviously, the product or service feature must be one the customer needs or desires. Moreover, such enhanced features and designs or advertising and marketing will increase costs, and customers must be price-insensitive — willing to pay for the differentiated product or service. This willingness to pay for the differentiated product or service is what provides the company relief from competitive pressure, specifically cost pressure. Firms pursuing a cost leadership strategy must make lower production and distribution costs their priority (Helmold *et al.*, 2022). By keeping their costs lower than those of their competitors through lean processes, enterprises using cost leadership can still price their products up to the level of their competitors and still maintain higher gross profit margins. Alternatively, these firms can price their products lower than those of their competitors in the hope of achieving greater market share and sales volume at the expense of gross profit margins. A focus strategy is based on a particular market, customer, product, or geography. A focus strategy is a concentrated and narrowly focused niche strategy (Slack & Brandon-Jones, 2021).

References

Ahmadin, C. & Lincoln, E. J. (2001). Keiretsu, governance, and learning: Case studies in change from the Japanese automotive industry. *Organization Science*, 12(6), 683–701.

Belekoukias, I., Garza-Reyes, J. A., & Kumar, V. (2014). The impact of lean methods and tools on the operational performance of manufacturing organisations. *International Journal of Production Research*, 52(18), 5346–5366.

Bertagnolli, F. (2020). *Lean Management. 2. Auflage. Einführung und Vertiefung in die japanische Man-agement-Philosophie*. Wiesbaden: Springer.

Helmold, M. & Terry, B. (2021). *Operations and Supply Management 4.0 Industry Insights, Case Studies and Best Practices*. Cham: Springer.

Helmold, M. *et al.* (2022). *Lean Management, Kaizen, Kata and Keiretsu. Best-Practice Examples and Industry Insights from Japanese Concepts*. Cham: Springer.

Johnson, G. *et al.* (2017). *Exploring Strategy* (11th edn.). Hoboken, New Jersey: FT Prentice Hall.

Liker, J. K. (2004). *The Toyota Way*. Madison: Mc Graw-Hill.

Ohno, T. (1990). *Toyota Production System. Beyond Large Scale Production*. New York: Productivity Press.

Porter, M. E. (1985). *Competitive Advantage. Creating and Sustaining Superior Performance*. New York: Free Press.

Slack, N. & Brandon-Jones, B. A. (2021). *Slack: Operations and Process Management. Principles and Practice for Strategic Impact* (6th edn.). London: Pearson.

Chapter 4

Lean and Learning Organization

4.1 Definition and Scope

A lean and learning organization can be defined as any organization that prioritizes personal and professional growth through knowledge transfer. These organizations encourage learning as part of their fundamental culture and overall vision for long-term success. A learning organization is a company that facilitates the learning of its members and continuously transforms itself (Ahmadin & Lincoln, 2001; Helmold & Terry, 2021).

4.2 3M-Concept to Become a Lean and Learning Organization

4.2.1 Strategic Need

Learning culture is a strategic need that most learning organizations aspire to create and maintain. It encourages and empowers the workforce to learn new skills that help them face the challenges in their roles. The same skillset helps them adapt to the evolving changes in their respective industries. The link between learning culture and continuous improvement in any organization is evident. Therefore, it requires constant cultivation, initiatives to gradually engage learners, and a learning platform that's available wherever and whenever employees need it.

4.2.2 3Ms as Foundation for the Lean and Learning Organization

Meaning, management, and measurement are the foundations for a learning culture that any organization needs to create. They are critical metrics to measure the progress of transforming your organization into a "learning organization" and an effective trigger for development and implementation.

These three levers are interconnected and overlap on many levels. With accrued time, carefully cultivated attitudes and commitments, including management procedures, will eventually provide a solid foundation for building a "learning organization." Such an organization can face intensifying competition, technology disruption, and talent demand/supply preferences (Figure 4.1).

4.2.3 Meaning

Meaning is the discerning concept of a learning organization. It involves defining the steps and/or applicable practices that organizations need to implement to create the organization of the future. Meaning requires clearer guidelines for implementation, accompanied by experiential advice rather than high aspirations. Additionally, they will require tools

Figure 4.1.   3M concept for a lean and learning organization.

that will help them assess the volume of their organizational learning culture and its effectiveness.

4.2.4 Management and Leadership

Management has the maximum influence on learning experiences by leading and forming behaviors that reinforce learning. One way to reinforce learning is by establishing an open and supportive environment that stimulates the exchange of ideas. Organizations can also signal the importance of spending time identifying challenges, transferring knowledge, and reflecting on each experiment. These are some approaches that will help organizations create and execute a robust learning process and practice. Additionally, it advocates a subtle shift in their agenda toward a commitment to learning rather than continuous improvement (Bertagnolli, 2020).

4.2.5 Measurement

As organizations aspire to such a transition, the maxim "if you can't measure it, you can't manage it" holds. Measurement deliberately assesses how learning experiences or cycles are integrated and could be translated to improved performance. There are various reinvented metrics to measure post-learning changes. However, it is traceable through three overlapping stages: cognitive, behavioral, and improved performance. Exposure to new ideas will eventually expand knowledge, and organizations will begin to think differently. This will lead to adopting new insights and altering behavioral traits accordingly. Thus, these changes will lead to measurable results that your organization is targeting, whether it's an increase in market share, optimized delivery, or enhanced quality. All of these will require constant learning culture audits to ensure that learning is serving the organization's needs (Belekoukias *et al.*, 2014).

4.3 Transformation to a Lean and Learning Organization — Kyoiku

A success factor in becoming a lean and learning organization is the concept of Kyoiku within the company culture and across the value chain (Helmold *et al.*, 2022). Kyouiku (教育) is the term for learning or education. Continuous improvement programs are proliferating as corporations seek to better themselves and gain an edge. Unfortunately, failed programs far outnumber successes, and improvement rates remain low. That's because most companies have failed to grasp a basic truth. Before people and companies can improve, they must first learn. And to do this, they need to look beyond rhetoric and high philosophy and focus on the fundamentals (Garvin, 1993).

Social education and lifelong learning (*shougai gakushuu*) are terms used to refer to companies, adult education, and related activities in Japan. Adult education originated immediately after the Second World War with the enactment of the Fundamental Law of Education (1947; revised in the 1980s). Since the mid-1980s, lifelong learning has been the more commonly used term. Lifelong learning involves not only going to school to learn how to read and write but also involves the learning that takes place outside the classroom through volunteering, outdoor activities, sporting events, and so on. Companies in Japan promote learning activities of their employees. Companies have become learning organizations. A learning organization is ideally a system that is constantly in motion. Events are taken as suggestions and used for development processes in order to adapt the knowledge base and scope for action to the new needs. This is based on an open and individualized organization that allows and supports innovative problem-solving. Mechanisms that support such learning processes are:

- Clear mission and visions with common target processes.
- Orientation of the organization toward the benefit of the customers.
- Ability to cooperate and resolve conflicts with mutual trust and team spirit.
- Process orientation and self-regulation in groups.
- New leadership style, support for new ideas (especially through leadership), idea management, integration of personnel, and organizational development.
- Rewarding commitment and tolerance for mistakes in risky endeavors.
- Ability to (self-) observation (well-functioning forecasting and communication systems — quicker, more precise, and able to give an overview of the effects of the most important processes).

4.4 The Learning and Improving Company

The successful transformation toward a lean and learning organization or company requires a modern mindset, which is incorporated into the corporate strategy and culture. Companies that follow this transformation path are called Gakushyu Kigyiou (学習企業). In order to be successful, organizations should learn and respond to changes quickly (Johnson *et al.*, 2017). They should learn to effectively challenge conventional wisdom, manage the organization's knowledge base, and make the desired changes. All organizational members take active part in identifying and resolving work-related issues. In a learning organization, employees practice knowledge management. They continuously acquire, share, and apply new knowledge in making decisions. In today's competitive world, organizations that learn and apply new concepts have an edge over competitors (Garvin, 1993). All organizations learn, whether they consciously choose to or not — it is a fundamental requirement for their sustained existence. People continually expand their capacity to create the results they truly desire, where new and expensive patterns are nurtured, where collective aspiration is set free and where people are continually learning how to learn together. Learning organization "facilitates the learning of all its members and continually transforms itself." Peter Senge, author of *The Fifth Discipline: The Art & Practice of the Learning Organisation*, popularized the term "Learning Organization" in the early 1990s. He defined learning organizations as "organizations that encourage adaptive and generative learning, encouraging their employees to think outside the box and work in conjunction with other employees to find the best answer to any problem" (Senge, 2006).

Senge is a supporter of decentralized and lean leadership, a model in which all people in an organization can work toward a common goal (Gilbert & Bower, 2002). His five disciplines of a learning organization outline how that can happen:

(1) **Personal Mastery:** In an interview, Senge called personal mastery the "cornerstone" of a learning organization. Personal mastery is the development of the capacity to accomplish personal goals; learning organizations make this possible by creating an environment where employees can, through reflection, develop their own sense of vision — how they look at the world, what matters to them, and what they are passionate about contributing to the world. Senge states, "Personal vision is the soil in which shared vision can be grown."

(2) **Shared Vision:** A shared vision is only possible in an environment of trust and collaboration instead of compliance with directives from above. Corporate leadership works together with employees toward a common vision — creating an environment where employees feel heard and are encouraged to take risks.

(3) **Mental Models:** With a mental model, we understand how our deeply ingrained assumptions and generalizations affect our interactions and decisions. To paraphrase Senge, understanding the difference between hearing what someone said, and truly understanding what they said, and understanding the gap between what actually happened and what we perceived happening requires reflection. "In a nonreflective environment, we take what we see as truth," argues Senge.

(4) **Team Learning:** Senge says that team learning can only happen when team members are "humble," when they are willing to reflect and take into account other people's views, suspending personal biases in order to work as a whole in a collaborative environment.

(5) **Systems Thinking:** Systems thinking is the idea that we're part of an interrelated system — not a disjointed set of personal silos; systems thinking addresses the whole and creates an understanding of how parts are interconnected. According to Senge, "[s]ystems thinking is a sensibility — for the subtle interconnectedness that gives living systems their unique character."

In lean management organizations, a sixth dimension has been added to the already existing five dimensions (Helmold *et al.*, 2022). This is continuous improvement, which is an important success factor for enterprises.

(6) **Continuous Improvement:** The organization must implement a culture and system of continuous learning throughout primary and secondary functions.

Figure 4.2 depicts the characteristics of a lean and learning organization.

4.5 Creating a Logical and Open Mind

Chiiku (知育) means to master intellectual knowledge and develop logical thinking for fundamental survival skills. For businesses to stay profitable, they first need stability based on a concrete understanding of their

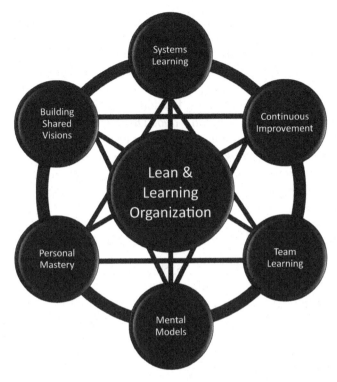

Figure 4.2.   Characteristics of a lean and learning organization.

needs and priorities. Then, by using their uncovered resources, they can begin to innovate. Understanding this fundamental need for the business's survival is the foundation of future prosperity, and it should also form the foundation for developing leaders within the workplace (Liker, 2004). Chiiku focuses on this logical understanding of the business in a larger context. This is like envisioning a forest as an entire ecosystem rather than just a collection of trees. For business leaders, *chiiku* means to calculate the sense of urgency and communicate it at all times (Ohno, 1990; Helmold, 2021).

4.6 Leadership Development and Culture

Tokuiku (德育) means to develop your rational interpersonal skills as a leader. Rational development (*tokuiku*) is different from logical development (*chiiku*). Logical thinking is based on cause and effect, whereas rational thinking is based on quantity and scale. Logical thinking can tell us what we ought to do, but we need to be rational to understand why it benefits each individual. It is necessary for humans to develop themselves by not staying satisfied with the current state and rationally comparing it to other possibilities. Leaders must first develop the courage to take risks and surpass the *status quo*. This is like ensuring the survival of the forest by understanding the needs of each organism that makes up the ecosystem. While *chiiku* is focused on the organization's survival as an entity, *tokuiku* focuses on ensuring the enrichment and success of the individuals who make up that entity (Helmold *et al.*, 2022).

4.7 Enhancing Physical and Mental Strength

Taiiku (体育): Modern education systems understand *taiiku* as physical education (PE). It is seen as simply a way to make students exercise their bodies through sports. But physical education is about more than just building muscles and developing a sense of competitiveness. Taiiku first focuses on strengthening one's willpower and emotions to elicit the right actions. For business leaders, *taiiku* means learning the skills to inspire a culture of immediate action, not just the words. Leaders must learn to help others break the *status quo*. This means learning the self-criticism mentality (*hansei*). Figure 4.3 shows a lean management workshop conducted by Dr. Marc Helmold within his organization in China. Enhancing the potential of employees physically and mentally through a continuous learning culture will lead to higher motivation and increased productivity across the value chain (Helmold *et al.*, 2022).

4.8 Learning from Experience

Senpai (先輩) and Kōhai (後輩) mean "senior" and "junior," and these hierarchies permeate all relationships in Japanese society, not just at work. A *senpai* is usually older and of higher ranking, or has been at the company longer. They usually act as a mentor for *kōhai*, although in reality not all *senpai* are as generous or genuine about mentoring junior colleagues. This hierarchy dictates all sorts of unspoken rules in Japanese work culture, including where you sit at the meeting table, how you conduct yourself at drinking parties, and what level of politeness you need to use

Figure 4.3. Lean management workshop.

Figure 4.4. Lean management and employee workshop.

(Sinha & Matharu, 2019). As an outsider to a Japanese company, all the etiquette surrounding *senpai* and *kōhai* culture can be bewildering, but it's something that you will pick up quickly once you're in that environment.

Figure 4.4 shows a team outing, lean management, and employee workshop at Bombardier in China. The overall goal was to learn from each other and to improve. The workshop was led by the general manager, Dr. Marc Helmold, accompanied by senior managers and facilitated by human resources.

References

Ahmadin, C. & Lincoln, E. J. (2001). Keiretsu, governance, and learning: Case studies in change from the Japanese automotive industry. *Organization Science*, 12(6), 683–701.

Belekoukias, I., Garza-Reyes, J. A., & Kumar, V. (2014). The impact of lean methods and tools on the operational performance of manufacturing organisations. *International Journal of Production Research*, 52(18), 5346–5366.

Bertagnolli, F. (2020). *Lean Management. 2. Auflage. Einführung und Vertiefung in die japanische Man-agement-Philosophie.* Wiesbaden: Springer.

Garvin, M. (1993). Organizational learning. Building a learning organization. *Harvard Business Review.* https://hbr.org/1993/07/building-a-learning-organization (Accessed February 1, 2024).

Gilbert, C. & Bower, J. L. (2002). Disruptive change. When trying harder is part of the problem. *Harvard Business Review*, 80(5), 94–101.

Helmold, M. (2021). *Kaizen, Lean Management und Digitalisierung. Mit den japanischen Konzepten Wettbewerbsvorteile für das Unternehmen erzielen.* Heidelberg: Springer.

Helmold, M. & Terry, B. (2021). *Operations and Supply Management 4.0 Industry Insights, Case Studies and Best Practices.* Cham: Springer.

Helmold, M. *et al.* (2022). *Lean Management, Kaizen, Kata and Keiretsu Best-Practice Examples and Industry Insights from Japanese Concepts.* Cham: Springer.

Johnson, G. *et al.* (2017). *Exploring Strategy* (11th edn.). New York: FT Prentice Hall.

Liker, J. K. (2004). The Toyota Way. Madison Mc Graw-Hill.

Ohno, T. (1990). *Toyota Production System. Beyond large Scale Production.* New York: Productivity Press.

Senge, P. (2006). *The Fifth Discipline (Rough Cut): The Art & Practice of the Learning Organization.* New York: Currency Publishing.

Chapter 5

Lean Leadership — Empowerment

5.1 Scope and Definition

Lean leadership is a management philosophy and approach that focuses on continuous improvement, waste reduction, and increased organizational efficiency. Lean leadership is characterized by the fact that, in addition to improving company processes (e.g., "avoiding waste," "increasing value creation"), the human factor is increasingly taken into account. The motto is: "Enabling instead of teaching." Lean leadership aims to gradually increase the ability of managers to develop their own competence and that of their employees, so that not only the acute problems are solved correctly but also the competence to continuously improve the performance of the entire organization increases (Bertagnolli, 2023). The origins of lean leadership lie in lean management and became known through the leadership development model at Toyota (Lean Leadership Development Model). The understanding of leadership in lean leadership and therefore also in Hoshin Kanri (policy deployment) differs greatly from conventional leadership models and management styles. Lean leadership pursues the overarching goal of establishing good problem-solving processes in the organization so that achieving goals becomes a natural process. Companies that practice lean leadership take a very systematic approach to identifying and developing their managers. They are based on a leadership development model that includes the following stages:

- *Develop yourself as a manager*: The (future) core competence of managers is the ability to reflect on their own behavior and work and to systematically increase their own "performance." That's why the first stage of leadership development is your own development.
- *Coaching and developing others*: The second stage is to take on the core task of developing other people as a manager — with the overarching goal that they in turn acquire the competence to reflect on their behavior and their work and to develop (which in turn relieves the managers).
- *Support daily Kaizen*: The first two levels mainly concern individual leadership. From the third level onward, the institution is the focus. This means that now it's about aligning groups of employees (teams, departments, areas) in one direction and ensuring that Kaizen is maintained and improved.
- *Create a vision and coordinate the goals*: Ideally, all managers and the entire organization are involved in the final development stage. It means that all activities are coordinated so that the challenging corporate goals are achieved (Hoshin Kanri). This requires that there is no silo thinking in the areas, that the (area) goals are coordinated and aimed at achieving the highest corporate goals, and that resources are used accordingly.

5.2 Empowerment: Key Success Factor in Lean Leadership

Lean leadership serves to convey meaning in daily work by showing the desired overarching corporate goals. Managers act as coaches and develop their employees' abilities to independently question and improve all processes and methods. Through their presence at the point of value creation, they communicatively involve employees in problem-solving and decision-making processes, thus enabling direct participation and a corporate culture that tolerates errors. As role models, they have a direct effect on the attitude of employees and increase their motivation. This interaction leads to a continuous improvement process (CIP), with which even challenging goals can be achieved.

5.3 Improvement and Coaching Kata as Triggers in Lean Leadership

5.3.1 Improvement Kata

The improvement *kata* (Figure 5.1) is a neutral learning routine that is independent of the given circumstances, which aims at a final state that corresponds to the corporate vision achieved via step by step and experimentally defined stage goals, so-called target states. The target states are derived from a well-founded understanding of the actual state (work steps, sequences and times, process properties, process indicators, and result indicators). The solution is not specified because it results from the procedure and can neither be forecast nor planned. The unknown solution is gradually developed through a regular examination of one's own actions and the given framework conditions and is only found at the end of the process.

The improvement *kata* is based on the following four elements:

- Understanding the vision and thus the direction of development.
- Recording the current situation.

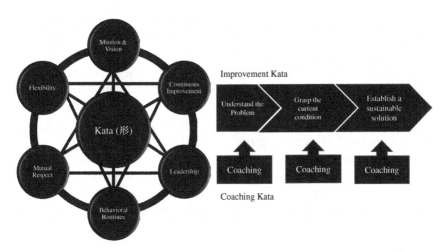

Figure 5.1.   Coaching and improvement *kata*.

- Determination of the next target state in each case.
- Step-by-step approach to the next target state (via the PDCA cycle). Obstacles provide indications of the need for action.

Independent striving for continuous improvement requires employees to consciously and openly examine their level of knowledge. They have to develop their own abilities and skills step by step in the sense of life-long learning and thus broaden their horizons. In doing so, they have to combine two thought systems of the human brain:

- The fast subconscious system that accesses stored knowledge via pattern recognition and leads to reflex-like actions.
- The slower conscious system that systematically pervades events and creates new insights and results through analysis and targeted linking.

The employees need support in the form of coaching *kata*.

5.3.2 Coaching Kata

The goal of coaching *kata* is to enable (direct) executives to support employees in developing their own systematic, experimental solutions on the way from the *status quo* via target states to the final state and let the employees applying the improvement *kata* to their routine. Coaching *kata* therefore does not provide a specific solution.

A manager, as a coach, therefore has the task of increasing the employees' confidence, self-confidence, and self-motivation so that they can meet and master growing challenges. As a prerequisite for the success of *kata* coaching, a trusting, mutually respectful relationship between manager and employee is essential.

Coaching *kata* is based on four questions, which are the starting point for approaching the desired target state. With this sequence, the coach trains the mental pattern and the (behavioral) routine of the employees in order to clarify and apply the improvement *kata* in every process or situation:

- What is the current state and what is the desired end state?
- What are the obstacles on the way to the final state?

- Which conclusions can be drawn from this and which next steps can be derived from it?
- By when can the first results be expected and what has been learned?

Coaching *kata* is practiced by many companies that want to anchor the lean concept and the principle of CIP, as it can also create the conditions for proven lean management methods (such as Kanban) to be beneficial because, in this case, employees have already internalized the pursuit of improvement.

5.4 Necessary Leadership Attributes

5.4.1 Scope and Definition

Successful organizations do not prosper by devoting a ruthless approach to chip away at costs, relentlessly reducing all decision-making to a reduction in headcount. The purpose of lean culture change is to secure the future of the entity by uniting its people to deliver the voice of the customer. In the not-for-profit sector, lean culture change drives organizational success by adding value to existing consumers and winning loyalty. It is about developing resilient service provision, developing core staff competencies, and attracting and retaining the best people. The lean culture reverses the polarity of the organization, shifting from a fire-fighting mode to a planning mode, where prevention of problems rather than reacting after the event of failure becomes the norm. Healthy, positive, and organizational cultures are characterized by a long-term continuity perspective, with a focus on tactics to resolve immediate short-term problems. The dominant culture should support and reward cross-organizational working. Implementing lean thinking is a cultural change that requires leadership because in the end, it's all about people. Here are 10 guidelines your leader can adopt right now to change the culture.

5.4.1.1 *Challenge People to Think*

If you are not thinking, you're not learning new things. If you're not learning, you're not growing — and over time becoming irrelevant in your work. The most successful leaders understand their colleagues' mindsets, capabilities, and areas for improvement. They use this knowledge/insight to challenge their teams to think and stretch them to reach for more.

5.4.1.2 *Lead by Example*

Leading by example sounds easy, but few leaders are consistent with this one. Successful leaders practice what they preach and are mindful of their actions. They know everyone is watching them and therefore are incredibly intuitive about detecting those who are observing their every move, waiting to detect a performance shortfall.

5.4.1.3 *Take Lots of Leaps of Faith*

Making a change requires a leap of faith. Taking that leap of faith is risky, and people will only take active steps toward the unknown if they genuinely believe — and perhaps, more importantly, feel — that the risks of standing still are greater than those of moving forward in a new direction. Making a change takes lots of leaps of faith.

5.4.1.4 *Create an Environment Where It is Acceptable to Fail*

Failure should be encouraged! That's right. If you don't try, you can't grow; and if growth is what you seek, failing is inevitable. There must be encouragement to try, and it's ok if you try and it doesn't work. An environment where you can't fail creates fear.

5.4.1.5 *Eliminate Concrete Heads*

"Concrete Heads" is the Japanese term for someone who does not accept that the organization must be focused on the elimination of waste. People feel threatened by the changes brought about by lean. As waste and bureaucracy are eliminated, some will find that little of what they have been doing is adding value. The anxiety they feel is normal and expected. To counteract this, it is critical that people are shown how the concept of work needs to change.

5.4.1.6 *Be a Great Teacher*

Successful leaders take the time to mentor their colleagues and make the investment to sponsor those who have proven they are able and eager to advance. They never stop teaching because they are so self-motivated to learn themselves.

5.4.1.7 *Show Respect to Everyone*

Everyone desires respect. Everyone. Regardless of your position or power, ensure you show respect to everyone. Everyone wants to be treated fairly.

5.4.1.8 *Motivate Your Followers*

Transformational leaders provide inspirational motivation to encourage their followers to get into action. Of course, being inspirational isn't always easy. Some ideas for leadership inspiration include being genuinely passionate about ideas or goals, helping followers feel included in the process, and offering recognition, praise, and rewards for people's accomplishments.

5.4.1.9 *Develop a True Team Environment*

Create an environment where working as a team is valued and encouraged and where individuals work together to solve problems and help move the organization forward. Individuals who will challenge each other and support each other make teams more successful.

5.4.1.10 *Encourage People to Make Contributions*

Let the members of your team know that you welcome their ideas. Leaders who encourage involvement from group members have been shown to bring about greater commitment, more creative problem-solving, and improved productivity. Constant change is a business reality and organizations must continually adapt to their environments to stay competitive or risk losing relevance and becoming obsolete. For each change, leaders must define it, create a vision of the post-change world, and mobilize their teams to make it. Fundamentally, a change of culture occurs when people start behaving differently as a result of a change in the climate of the organization (Helmold *et al.*, 2022). There are many different models of how an organizational culture is shaped by the prevailing climate and how it can be assessed. Leaders who protect the *status quo* through control must surrender to change in order to secure the future of their organization. Don't be the leader who rewards herd mentality, and "me too" thinking. Don't be the leader who encourages people not to fail or not to take risks. Be the leader who both models and gives permission to do the exact opposite of the aforementioned — be a leader who leads. The culture of an organization is learned over time. It can be taught to new employees through formal training programs but is more generally absorbed through stories, myths, rituals, and shared behaviors within teams. Organizational culture will have a positive or negative impact on everything you try to do whether you want it to or not.

Aligning a company according to proven lean principles means creating more agile, faster, and significantly cheaper processes in the face of ever-increasing competition. However, lean projects only work if employees are included in the entire lean process, motivated, coached, and empowered to take personal responsibility. In all lean projects, leadership is the decisive factor (Rodermund, 2021).

5.5 Striving for Personal Mastery

5.5.1 Scope and Definition

The Shu Ha Ri concept has been derived from ancient Japanese martial arts and details the phases for achieving personal mastery (Liker & Convis, 2011). There are three stages of Shu Ha Ri that a novice needs to follow.

5.5.2 Shu — Discipline

Shu — The disciple first gets introduced to the traditional methods. He practices those techniques without having any second thoughts. It's just about following the instructions of the teacher and remembering each practice or ceremony thoroughly — things like what's the primary solution for a certain obstacle, which move you make in which condition, etc. Here, the student doesn't have to think about the hidden theory behind the methodology. In the case of complexities, the student is advised to follow one technique after another.

5.5.3 Ha — Combining Theory and Practice

Ha — After the basic practice, the student starts to pay attention to the theory behind the practice and assimilate his learnings with the practice. This is a stage which each and every one of us goes through at some point of time, and we all start questioning ourselves or the day-to-day behavior we are practicing.

It's also the stage where the student goes beyond the illusions by separating right from wrong by his own decision-making. Even sometimes, he learns different things from other resources and tries to understand the relations and differences deeply.

5.5.4 Ri — Becoming a Master

Ri — Now, the student doesn't need the practice to be taught by anyone. He understands it perfectly and learns the lessons from his own experience. He thinks beyond the lessons given by the teachers and tries to come up with methods that are his own and new for others. He surpasses his previous generation. Now, he is in a stage where he can be a coach or mentor for his colleagues or the upcoming generation.

5.6 Active Two-Way Communication — HoRenSo

Ho-Ren-So (Japanese: (報 連 相) is a mix of bottom-up and top-down communication between management and employees. It is not only used to solve problems but also to create transparency and promote the exchange of information. Because for us, there is no either/or when it comes to quality. Ho-Ren-So is an acronym and is made up of three words that also describe the process chain.

- Hokoku: Report the problem — immediately, precisely, objectively.
- Renraku: Contact the next level — open, fearless, trusting.
- Sodan: Consult with the management team and be in close communication with your employees in order to solve the problem quickly.

References

Bertagnolli, F. (2023). *Lean Empowerment. Die konsequente Fortsetzung von Lean Leadership*. München: Schäffer-Poeschel.

Helmold, M. *et al.* (2022). *Lean Management, Kaizen, Kata and Keiretsu: Best-Practice Examples and Industry Insights from Japanese Concepts*. Cham: Springer.

Liker, J. & Convis, G. L. (2011). *The Toyota Way to Lean Leadership: Achieving and Sustaining Excellence through Leadership Development*. New York: McGrawhill Education.

Rodermund, M. (2021). *Erfolgsfaktor Lean Leadership: Wege zu flexiblen und effizienten Prozessen*. München: Schaeffer-Poeschel.

Chapter 6

Waste and Value-added

6.1 Scope and Definition: Identification and Elimination of Waste

The fundamental and overriding principle of lean management is the identification of waste (Japanese: Muda, 無駄) and its continual elimination (Ohno, 1990). Elimination of waste and replacement by value or added value is therefore the major goal in enterprises and organizations (Liker, 2004).

Added value can be defined as products, services, processes, and activities which generate a certain value for the organization or enterprise. Value-added must be regarded from the customer's viewpoint and is everything for which the customer is willing to pay. It is important that value-added is recognized and perceived as value by the client (Bertagnolli, 2020). Many studies have shown that we only add value to a product for less than 5%–15% of the time, the rest of the time is wasted (Helmold & Terry, 2021). The opposite is non-adding value or waste. Waste is anything which adds cost or time without adding any value or any activity which does not satisfy any of the above conditions of "value-added is a waste or a non-value-adding activity in a process." The focus in operations management must therefore be on eliminating such activities as waiting time or rework (Liker, 2004). Enterprises must target value-added processes and eliminate or reduce waste, whereby waste can be visible (obvious) or invisible (hidden), as shown in Figure 6.1 (Helmold *et al.*, 2022). The main idea of lean management is to highlight the things that add value by reducing or eliminating everything else (waste) (Sahoo,

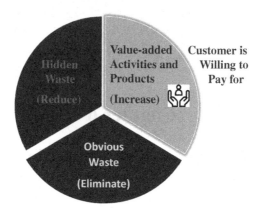

Figure 6.1.   Value added, obvious, and hidden waste.

Category	Impact	Principle
Value add	• Added value for products and services • Customer pays for it • Customer recognizes this as value add(ed)	**Increase, Improve**
Obvious Waste	• No added value for products or services • Tasks are visible and not necessary for production, but cannot seen without proper analysis	**Identify, Reduce, Eliminate**
Hidden Waste	• No added value for products or services • Inefficiencies are not visible, but tasks are necessary for production (without improvement)	**Identify, Reduce, Minimize**

Figure 6.2.   How to handle value added, obvious, and hidden waste.

2019). As a proven consequence, when you eliminate waste, the quality of products improves, while production time and costs are reduced. Figure 6.2 illustrates that waste must be ideally eliminated or reduced.

6.2   Concentration on Value-added Elements, Identification, and Elimination of Waste across the Value Chain

Lean management concentrates on customer satisfaction and value-added elements, as the customer is willing to pay for them. The major goal of the lean management philosophy is the identification and continuous elimination of waste across the value chain. Figure 6.2 shows how lean management enterprises handle value-added activities, obvious or hidden waste. Value-added products, processes, or elements must be improved, increased, and enriched, as the customer is willing to pay. On the contrary, waste is something for which the customer is not willing to pay. Therefore, it is necessary to eliminate and reduce the inefficiencies, which can be transparently identified and seen (obvious waste) (Bertagnolli, 2020). Hidden waste, bad products, or inefficient processes, which cannot be seen at first glance, may occur because tasks or activities may be necessary in the value chain without significant improvement and corrective actions. In this case, a profound analysis may be required for identification and systematic corrective actions to replace this type of waste by value-added (Helmold *et al.*, 2022).

6.3   Value-added beyond the Own Value Chain

The value chain of a company is a set of activities that a firm operating in a specific industry performs in order to deliver a valuable product or service to the market. The value chain integrates internal and external activities, as shown in Figure 6.3. The value chain is a supply chain framework which includes areas like suppliers, logistics, operations, marketing, sales, and quality as primary or secondary activities (Aberdeen Group, 2006). Primary activities contribute directly to the production of goods or services (procurement, production, marketing), while secondary functions (human resources, finance) support the primary functions (Bothard *et al.*, 2009). Companies should concentrate on their core competencies and reduce their value-added activities to 20% or less (Büsch, 2019). Core competencies are the origin of sustainable and long-term competitive advantages through superior manufacturing knowledge, faster processes, improved procedures, better quality, a certain brand name, efficiencies, innovations, patents, or advanced workforces. As many companies have external value chains (purchase of goods and services) of more than 80%,

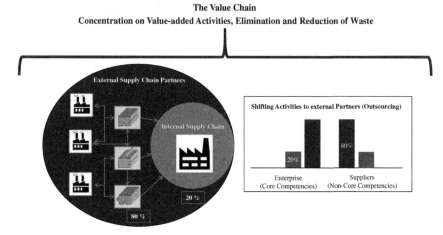

Figure 6.3.   Value added and the obvious and hidden waste across the value chain.

the procurement and supply function has an integral role in the enterprise of applying lean management elements toward the main suppliers (Adenso-Diaz *et al.*, 2012).

6.4   The Concept of Muda, Muri, and Mura

The Toyota Production System, and later on the concept of lean, was developed around eliminating the three types of deviations that show inefficient allocation of resources. The three types are Muda (無駄) (waste), Muri (無理) (overburden), and Mura (斑) (unevenness) (Figure 6.4). Muda means wastefulness, uselessness, and futility, which contradicts value-addition. Value-added work and activities are processes that add value to the product or service that the customer is willing to pay for. There are seven categories of waste under Muda Type 2 that follow the abbreviation TIMWOOD. The seven wastes are (1) Transport, i.e., excess movement of product, (2) Inventory, i.e., stocks of goods and raw materials, (3) Motion, i.e., excess movement of machine or people, (4) Waiting, (5) Overproduction, (6) Overprocessing, and (7) Defects. Muri means overburden, beyond one's power, excessiveness, impossibility, or unreasonableness. Muri can result from Mura and in some cases be caused by excessive removal of Muda (waste) from the process. Muri also exists when machines or operators are utilized for more than 100% capability to

Figure 6.4.   The concept of Muda (無駄), Muri (無理), and Mura (斑).

complete a task or in an unsustainable way. Muri over a period of time can result in employee absenteeism, illness, and breakdowns of machines. Standardizing work can help avoid Muri by designing the work processes to evenly distribute the workload and not overburden any particular employee or equipment. Mura means unevenness, non-uniformity, and irregularity. Mura is the reason for the existence of any of the seven wastes. In other words, Mura drives and leads to Muda. For example, in a manufacturing line, products need to pass through several workstations during the assembly process. When the capacity of one station is greater than that of the other stations, you will see an accumulation of waste in the form of overproduction, waiting, etc. The goal of a lean production system is to level out the workload so that there is no unevenness or waste accumulation. Figure 6.3 shows the elements Muda, Muri, and Mura.

6.5 Ishikawa Diagram to Identify Waste

Ishikawa diagrams (also called fishbone diagrams, herringbone diagrams, cause-and-effect diagrams, or Fishikawa) are causal diagrams created by Kaoru Ishikawa (Japanese: 石川 馨 Ishikawa Kaoru, 1915–1989) that show the cause–effect situation of a specific event. Common uses of the Ishikawa diagram are areas of design, supply, production, and quality defect prevention to identify potential factors causing an overall effect.

Each cause or reason for imperfection is a source of variation. Causes are usually grouped into major categories to identify and classify these sources of variation. The target of value-add and quality is shown as the fish's head, facing to the right, with the causes extending to the left as fishbones; the ribs branch off the backbone for major causes, with sub-branches for root causes, to as many levels as required. Figures 6.5 and 6.6 show two examples of the Ishikawa diagram. The advantage of using this method is the application of a highly visual brainstorming tool, which can spark further examples of the root causes. It serves to quickly identify if the root cause is found multiple times in the same or different causal

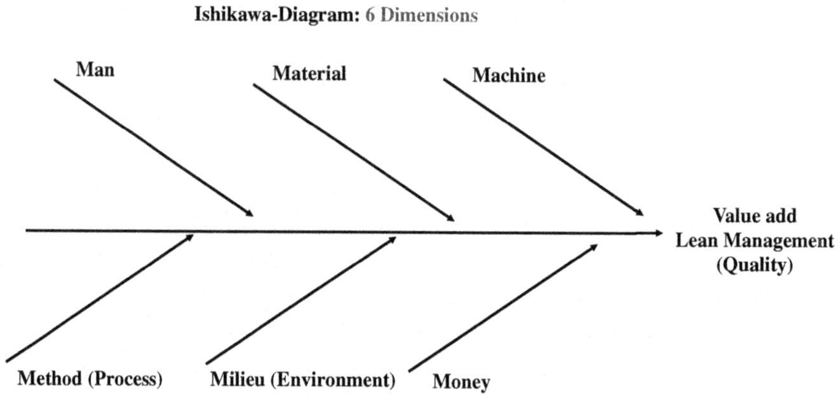

Ishikawa-Diagram: 6 Dimensions

Figure 6.5.   Ishikawa diagram — Six dimensions.

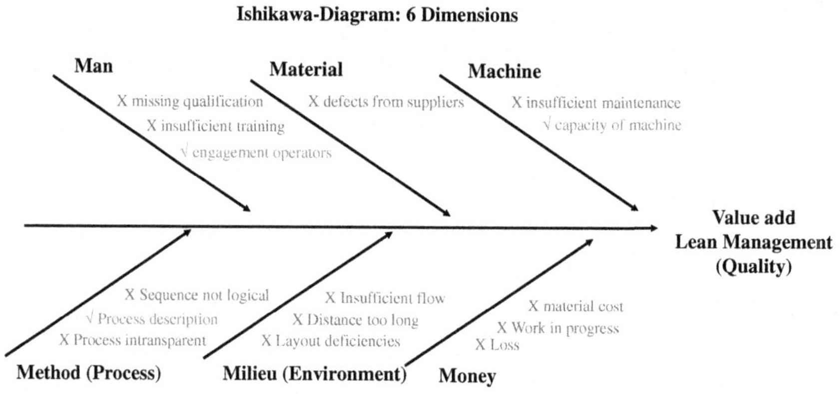

Ishikawa-Diagram: 6 Dimensions

Figure 6.6.   Ishikawa diagram — Practical example.

tree. The Ishikawa diagram is also a good visualization tool for presenting issues to stakeholders.

6.6 Eight Types of Waste: DOWNTIME Concept

6.6.1 Definition and Scope

The lean management framework was originally developed by Taiichi Ohno for Toyota and is utilized by companies across the globe. By utilizing the concepts of lean, organizations can begin to remove wastes from their processes while making better use of their talent and resources. The original seven wastes identified by Ohno have expanded to include one more related to an organization's employees (non-utilized talent) (Helmold & Terry, 2021). Enterprises can easily recall the eight wastes of lean by using the acronym DOWNTIME, which stands for:

1. Defects
2. Overproduction
3. Waiting
4. Non-Utilized Talent
5. Transportation
6. Inventory
7. Motion
8. Excess-Processing

6.6.2 Defects

Defects, as shown in Figure 6.7, refer to a product deviating from the standards of its design or from the customer's expectations. Defective products must be replaced; they require paperwork and human labor to process; they might potentially lose customers; the resources put into the defective product are wasted because the product is not used. Moreover, a defective product implies waste at other levels that may have led to the defect to begin with; making a more efficient production system reduces defects and increases the resources needed to address them in the first place. Environmental costs of defects are the raw materials consumed, the defective parts of the product requiring disposal or recycling (which wastes other resources involved in repurposing it), and the extra space required and increased energy use involved in dealing with the defects.

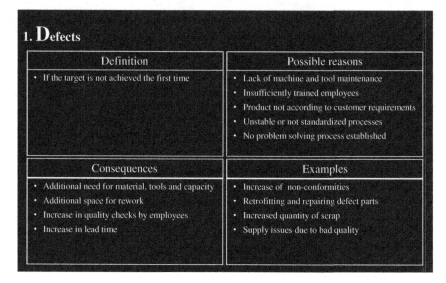

Figure 6.7.   Defects.

6.6.3 Overproduction

Overproduction is defined as any type of production that is more than needed or before it's needed. Overproduction is the situation in which too much material is produced due to waste (Helmold *et al.*, 2022). This can imply raw materials, semi-finished products, finished products, or assemblies. It can result in excess inventory, and it consumes time, effort, money, materials, and resources that could have otherwise been spent more effectively elsewhere (see Figure 6.8).

6.6.4 Waiting

Waiting is any idle or waiting time within the value chain. This waste comprises the idle time of operators or other employees in operations, and waiting for work to arrive or to be told what to do is a significant waste. Waiting or standstill times must be avoided, as waiting results in reduced efficiency and productivity. Other outcomes are longer lead times and decreasing engagement and motivation of employees, as illustrated in Figure 6.9.

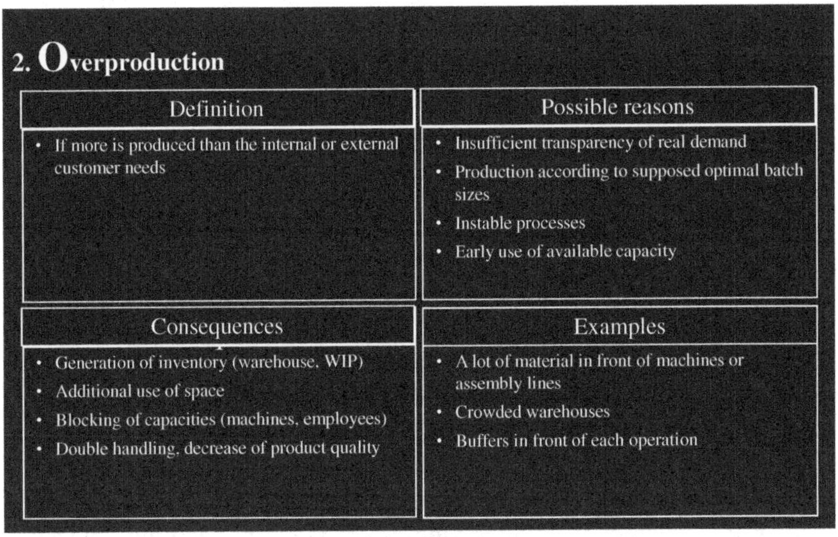

2. Overproduction

Definition	Possible reasons
• If more is produced than the internal or external customer needs	• Insufficient transparency of real demand • Production according to supposed optimal batch sizes • Instable processes • Early use of available capacity

Consequences	Examples
• Generation of inventory (warehouse, WIP) • Additional use of space • Blocking of capacities (machines, employees) • Double handling, decrease of product quality	• A lot of material in front of machines or assembly lines • Crowded warehouses • Buffers in front of each operation

Figure 6.8.   Overproduction.

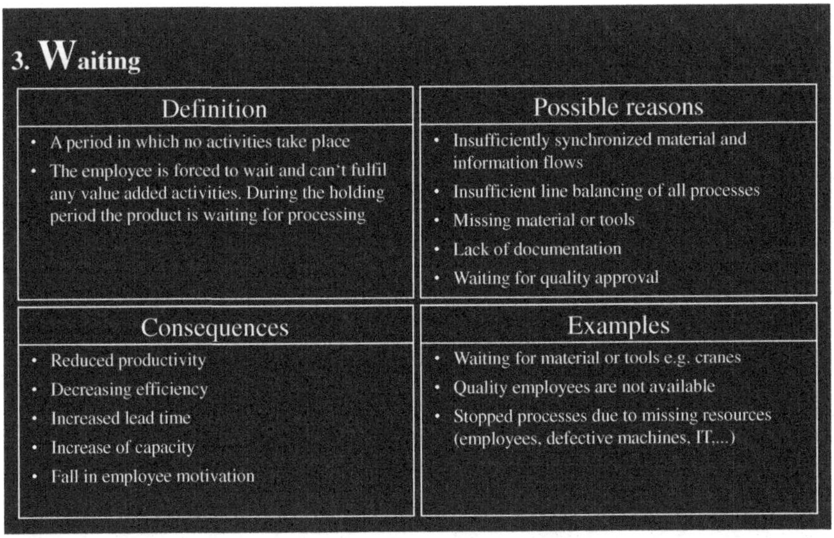

3. Waiting

Definition	Possible reasons
• A period in which no activities take place • The employee is forced to wait and can't fulfil any value added activities. During the holding period the product is waiting for processing	• Insufficiently synchronized material and information flows • Insufficient line balancing of all processes • Missing material or tools • Lack of documentation • Waiting for quality approval

Consequences	Examples
• Reduced productivity • Decreasing efficiency • Increased lead time • Increase of capacity • Fall in employee motivation	• Waiting for material or tools e.g. cranes • Quality employees are not available • Stopped processes due to missing resources (employees, defective machines, IT,...)

Figure 6.9.   Waiting.

6.6.5 Non-Utilized Talent

Non-utilized talent describes the missing use of employees and their potential. Not fully utilizing the experience, skills, knowledge, or creativity of employees will lead to demotivation, bad quality, and inefficiencies, as shown in Figure 6.10.

6.6.6 Transportation

Transportation throughout the own value chain, logistics, and supplier activities is a significant area of waste. Excess transportation is a significant waste because the time, manpower, energy, efforts, and resources required to move items are something the customer does not care about and does not want to pay for (Ohno, 1990). Examples of wastes of transport are the transport of products from one functional area, such as pressing, to another area, such as welding, or the use of material handling devices to move batches of material from one machine to another within a work cell. It wastes time because operators are dedicating the available time of the work day to moving items from one place to another. It wastes energy and resources in the sense that employee time could be better utilized and

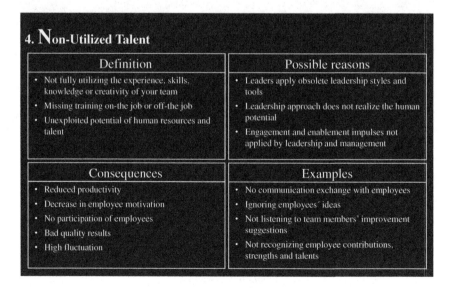

Figure 6.10.   Non-utilized talents.

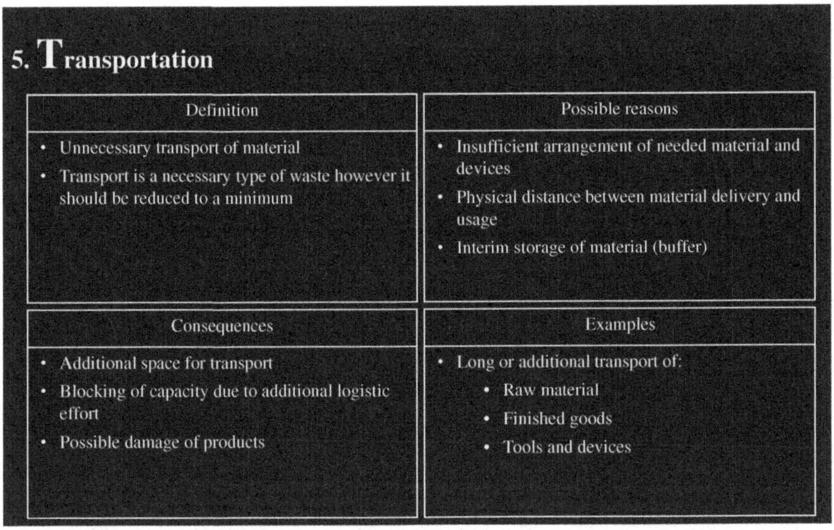

5. Transportation

Definition	Possible reasons
• Unnecessary transport of material • Transport is a necessary type of waste however it should be reduced to a minimum	• Insufficient arrangement of needed material and devices • Physical distance between material delivery and usage • Interim storage of material (buffer)

Consequences	Examples

• Additional space for transport
• Blocking of capacity due to additional logistic effort
• Possible damage of products
	• Long or additional transport of:
	  • Raw material
	  • Finished goods
	  • Tools and devices

Figure 6.11.   Transportation.

because some tools used for transportation (forklifts, trucks, pallet jacks) consume energy like electricity or propane. Also, by dedicating machines and operators' time to waste activities, they are no longer free and available to take on value-added activities (Sinha & Matharu, 2020). Figure 6.11 shows transportation waste. Reasons can include insufficient layouts and long distances between individual operations. The consequences of this waste are the increased time requirements and the decreased productivity (Helmold *et al.*, 2022). Decreased productivity will result in higher operating cost and can harm the profitability of the enterprise (Pinto *et al.*, 2018).

6.6.7 Inventory

Inventory and stored items are a significant waste and consist of excess goods, materials, or semi-finished goods. Finished or semi-finished goods inventory is generally the most expensive inventory as it has labor and other overhead attached to it along with the cost of material consumed during production. In order to reduce this inventory, process improvements as well as a higher accuracy in forecasting customer requirements are required. Inventory waste refers to the waste produced by unprocessed inventory. This includes the waste of storage, the waste of capital tied up

6. Inventory

Definition	Possible reasons

• More material than needed according to planning in terms of:
  • Raw material
  • Semi-finished parts
  • Work in progress (WIP)
  • Finished goods
	• Problems regarding planning and logistic processes
	• Bad supplier delivery performance and quality
	• High product variety

Consequences	Examples
• Capital costs and fixed assets expenses • Double handling, possible damages based on double handling, rework • Genuine problems won't be discovered and therefore not solved • Search effort • Scrap and obsolescences	• Overfilled warehouses • Overfilled place in production areas • Buffer stocks in production • Crammed corridors • Crammed desks

Figure 6.12.   Inventory.

in unprocessed inventory, the waste of transporting the inventory, the containers used to hold inventory, the lighting of the storage space, etc. Moreover, having excess inventory can hide the original wastes of producing said inventory. The environmental impacts of inventory waste are packaging, deterioration, or damage to work-in-process, additional materials to replace damaged or obsolete inventory, and the energy to light, as well as either heat or cool, inventory space. Figure 6.12 displays the definition, reasons, consequences, and examples for inventory. Inventory will have a negative impact on working capital and on cash flow, so sophisticated production planning must focus on the optimum levels of inventory throughout the value chain and operations (Helmold *et al.*, 2022).

6.6.8 Motion

Motion waste is the excessive movement of man, material, or machines within the workspace. Motion waste will lead to higher costs as productivity decreases. Another problem of motion is the necessity for more time and capacity in operations than is actually required. A proper workflow analysis and value stream mapping help to minimize this waste. Figure 6.13 outlines the definition, possible reasons, consequences, and examples of this waste.

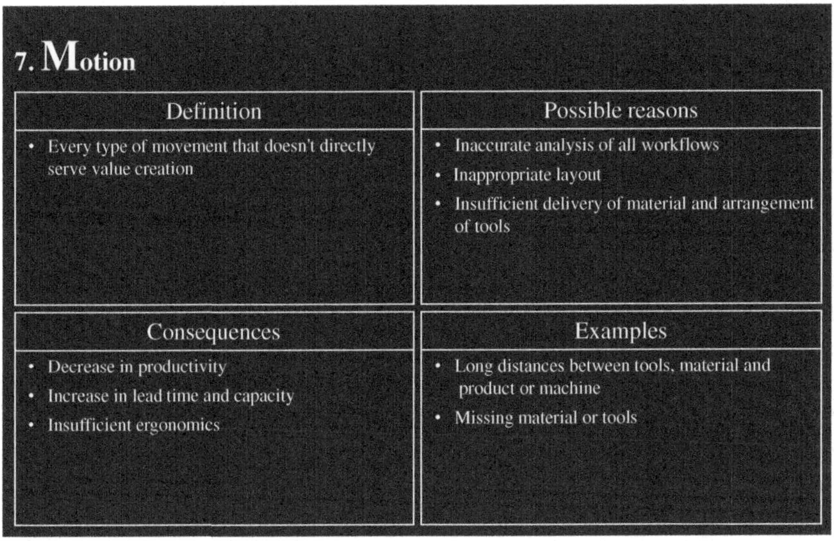

7. Motion

Definition	Possible reasons
• Every type of movement that doesn't directly serve value creation	• Inaccurate analysis of all workflows • Inappropriate layout • Insufficient delivery of material and arrangement of tools

Consequences	Examples
• Decrease in productivity • Increase in lead time and capacity • Insufficient ergonomics	• Long distances between tools, material and product or machine • Missing material or tools

Figure 6.13.   Motion.

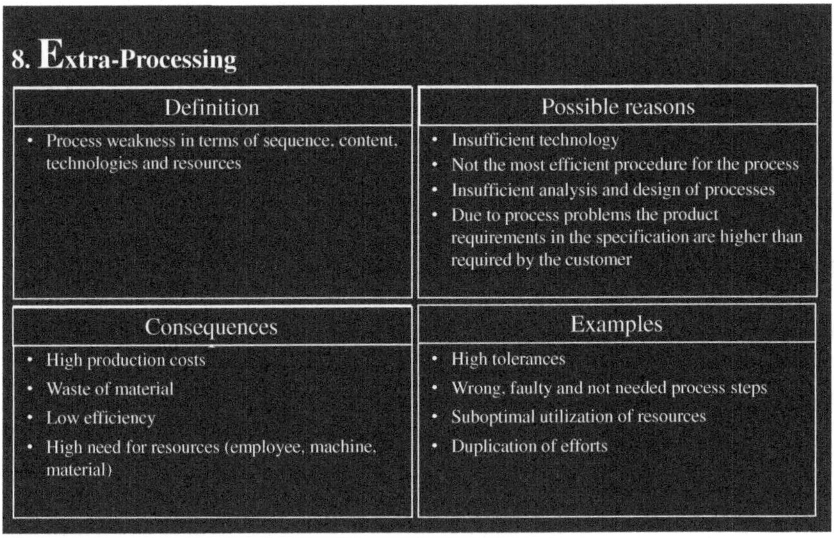

8. Extra-Processing

Definition	Possible reasons
• Process weakness in terms of sequence, content, technologies and resources	• Insufficient technology • Not the most efficient procedure for the process • Insufficient analysis and design of processes • Due to process problems the product requirements in the specification are higher than required by the customer

Consequences	Examples
• High production costs • Waste of material • Low efficiency • High need for resources (employee, machine, material)	• High tolerances • Wrong, faulty and not needed process steps • Suboptimal utilization of resources • Duplication of efforts

Figure 6.14.   Excess.

6.6.9 Excess-Processing

Excess-processing is related to all activities and processes in operations which are more than the customer really needs. Figure 6.14 highlights possible reasons for excessive processing, such as insufficient technology, bad design, inefficiencies, or unawareness of customer-specific requirements. Over-processing refers to any component of the process of manufacture that is unnecessary. Painting an area that will never be seen or adding features that will not be used are examples of over-processing. Essentially, it refers to adding more value than the customer requires. The environmental impact involves the excess of parts, labor, and raw materials consumed in production. Time, energy, and emissions are wasted when they are used to produce something that is unnecessary in a product; simplification and efficiency reduce these wastes and benefit the company and the environment (Niemann *et al.*, 2021).

References

Bertagnolli, F. (2020). *Lean Management. 2. Auflage. Einführung und Vertiefung in die japanische Management-Philosophie.* Wiesbaden: Springer.

Aberdeen Group. (2005). *Assuring Supply and Mitigating Risks in an Uncertain Economy. Supply Risk Management Benchmark.* Boston.

Adenso-Diaz, B., Mena, C. H., Garcia, S., & Liechty, M. (2012). Supply chain management: The impact of supply network characteristics on reliability. *An International Journal, 17*(3), 1–36.

Bothard, C. C., Warsing, D. P., Flynn, B. B., & Flynn, E. J. (2009). The impact of supply chain complexity on manufacturing plant performance. *Journal of Operations Management, 27*, 78–93.

Büsch, M. (2019). *Fahrplan zur Transformation des Einkaufs.* Wiesbaden: Springer.

Helmold, M. & Terry, B. (2021). *Operations and Supply Management 4.0: Industry Insights, Case Studies and Best Practices.* Cham: Springer.

Helmold, M. *et al.* (2022). *Lean Management, Kaizen, Kata and Keiretsu: Best-Practice Examples and Industry Insights from Japanese Concepts.* Cham: Springer.

Liker, J. K. (2004). *The Toyota Way.* Madison: McGraw-Hill.

Niemann, J., Reich, B., & Stöhr, C. (2021). *Lean Six Sigma. Methoden zur Produktionsoptimierung.* Cham: Springer.

Ohno, T. (1990). *Toyota Production System. Beyond Large Scale Production.* New York: Productivity Press.

Pinto, J. L. *et al.* (2018). *Just in Time Factory. Implementation Through Lean Manufacturing Tools*. Cham: Springer.

Sahoo, S. (2019). Lean manufacturing practices and performance: The role of social and technical factors. *International Journal of Quality & Reliability Management*, 37(5), 732–754.

Sinha, N. & Matharu, M. (2019). A comprehensive insight into lean management: Literature review and trends. *Journal of Industrial Engineering and Management*, 12(2).

Chapter 7

Agility and Lean Management as Success Factors

7.1 Scope and Definition

Lean management is a management and improvement methodology aimed at eliminating wasted time and resources through systematic analysis of processes and value streams. Agile here is an umbrella term for a philosophical approach to improvements, development, or processes across the value chain which prioritizes early and continuous delivery of valuable functionality that satisfies customers (Helmold, 2021).

Lean agile, or agile lean, is an agile methodology that, in basic terms, is quite simple: improve efficiency by eliminating waste. Unlike traditional, waterfall project management, which dictates a set plan laid out by a project manager, lean agile strives to reduce all tasks and activities that don't provide real value. This helps ensure everyone involved in a project or product development can work at optimal efficiency (Pendleton *et al.*, 2021).

7.2 Agility is a Competitive Driver for Success

Agility (Latin: *agilis*: nimble, agile) describes a modern form of work organization whose goal is, in particular, to achieve flexibility, adaptability, and rapid development in short iterative cycles. Agile and cross-departmental teams focus on quality and value creation at a very early stage of the process, as necessary changes can be identified and implemented at an early stage, and projects can therefore be introduced more quickly and flexibly. Processes and procedures are deliberately kept lean (Helmold, 2022). An agile company strives to make future developments part of everyday work. The internal organization is constantly changing. This means the company is able to adapt quickly and make better use of opportunities (Neun, 2020). Regular communication and transparency are central points of agile organizations. Groups and teams have a high degree of personal responsibility, self-management, and freedom of choice. An agile culture is an important prerequisite for the successful introduction of agile methods. Agile organization is particularly useful in complex environments, as unexpected changes often occur that require changes in planning. The use of agile methods with a focus on quality and value creation is becoming increasingly relevant for companies in order to secure long-term competitive advantages (Peters, 2015). With flexible and lean development processes, customer products and services can be developed faster and in a more customer-oriented manner, which optimizes development and start-up costs and shortens the time to market. The aim is to deliver executable prototypes at an early stage, to receive quick and regular feedback from the customer, and to use this for further work.

7.3 Key Success Factors for Becoming a Lean and Digital Company: Agility and Leadership

7.3.1 Scope and Definition

Companies must be agile and innovative to be successful in the face of constant change. Agility makes it possible to react to changes more quickly with flexible approaches and services (Wittenstein, 2022). Agility and lean process management are important factors for the successful implementation of ideas and innovations (Knuppertz & Ahlrichs, 2022). Agility is the ability of a company to act flexibly and proactively in order to introduce necessary changes and innovations (Helmold, 2022). Management plays a key role in this context, as it must integrate and motivate all employees into the concept using an agile leadership approach (Fatma, 2015). A corporate culture with agile ideas creates the conditions for teams and employees to better manage complex requirements and be more innovative (Helmold & Samara, 2019; Mollenhauer & Sommerlatte, 2016). Agile ideas and measures can be tried out and successfully introduced in the form of both small and complex projects,

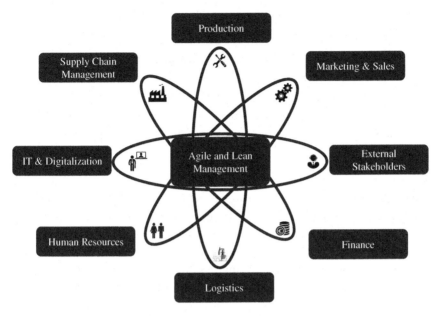

Figure 7.1. Agile organizations.

completely independent of the respective industry and the size of the company (Hofert, 2022). Figure 7.1 shows how agility can be successfully implemented in companies. Agile working avoids rigid hierarchies and aims for quick results, which are improved quickly and frequently through close coordination with the (internal or external) stakeholders from different departments. For example, the agile framework Scrum only describes three roles. The first role includes the Scrum Master, who ensures that the procedure is applied correctly. The second role is played by the product owners, who represent the interests of the customer and client. The final role is played by team members or people on the development team. This team is responsible for the planning, division of work, implementation, progress, and results of the project. Agility essentially has six dimensions (Table 7.1).

7.3.2 Mission and Vision as Agile Paradigm

Many companies are starting agile change processes in their digitalization projects. This requires a sustainable and transparent mission and vision.

Table 7.1.   Six dimensions for agility.

Dimension 1	Agile mission, vision, and paradigm
Dimension 2	Customer-focused and targeted project structure
Dimension 3	Iterative process layout and continual improvement processes
Dimension 4	Employee-focused leadership behavior
Dimension 5	Agile leadership tools and methodologies
Dimension 6	Agile and lean organizational structure and set-up

After some time, the question arises as to whether agility should only be understood at the project or product development level or for other areas in the company. Basically, a distinction can be made between two types of agility (Helmold, 2022). Corporate agility refers to agile collaboration for large-scale products that usually involve many different teams. Business agility refers to the agility of the entire organization and includes all areas, not just product development.

7.3.3 Customer-Focused and Targeted Project Structure

Traditional organizations focus very much on themselves. They think in pyramids and silos. Agile companies, on the other hand, align their strategy with the customer and strive to maximize customer benefit. Agile customer-oriented organizations are characterized by network structures with internal and external stakeholders instead of hierarchies. The focus is on the team-based process organization instead of the non-value-adding organizational structure.

7.3.4 Iterative Process Layout and Continual Improvement Processes

Agile companies rely on an iterative approach and delivering in increments, i.e., short-term results. Agile customer-oriented organizations plan their processes, products, and services iteratively instead of following the waterfall model. This reduces the time required for planning and conception. Customers receive the products and services in quick succession in smaller parts instead of in one piece over a longer period of time. Agile processes are iterative and incremental. They focus on short-term results and enable rapid adaptation to changing conditions. Errors are identified early and can be corrected promptly.

7.3.5 Employee-Focused Leadership Behavior

In agile companies, managers put themselves at the service of the teams in order to create benefits and added value for the customer more quickly. In agile organizations, managers are not controlling superiors who put pressure on their employees, but rather they delegate responsibility to the employee teams. Therefore, plural leadership approaches continue to gain importance (Burns, 2018).

7.3.6 Agile Leadership Tools and Methodologies

Human resources (HR) works in dialogue with employees and managers and generates values with clear customer benefits. HR is the decisive catalyst for agile transformation. In agile organizations, employees are heavily involved in personnel planning. Employee development does not (only) take place on the basis of specifications but (also) within the teams themselves (peer feedback).

7.3.7 Agile and Lean Organizational Structure and Set-up

Agile organizational cultures are characterized by transparency, dialogue, an attitude of trust, and short-term feedback mechanisms. In "classically" organized structures, a culture of strict rules, standardized specifications, and little freedom of choice for employees often prevails. In agile organizations, knowledge is passed on openly, errors are addressed openly and constructively, and status symbols (executive floor, carpet floor) are eliminated.

7.4 Agile and Lean Leadership

Agile and lean leadership means that the company can adapt to changes as quickly as possible. It supports employees at eye level in finding the best solutions to challenges together. The needs of all stakeholders are taken into account. The central values of agile leadership are openness to new things, communication, and flat hierarchies (Scherer, 2018). Agile companies are able to align their organization and business model to new market requirements in a short time. In addition, agile organizations are proactive in seizing opportunities that arise. This applies to all areas of the company. Agility in the company is the right mix of "doing agile" (methods) and "being agile" (mindset). In the course of digital transformation, agility is a necessary leadership and organizational principle in order to be successful in digital markets (Pfeiffer, 2021). Agile transformation is becoming an essential cornerstone of a successful digital transformation. Figure 7.1 shows a prototype of the agile organization. Agile companies have their own DNA. It is difficult to attribute their success to individual measures or methods. Rather, their success is based on six pillars: as mission, customer focus, agile and (re)thought leadership, agile methods, continuous improvement, and an agile corporate culture (Gloger, 2014). In agile organizations, the customer is part of value creation and not simply the last link in the chain. Requirements are formulated in user stories, and value creation is thought of starting from the customer and along his customer journey instead of from the inside out (Helmold *et al.*, 2020). An ideal and impressive symbolism for the customer's position is to give him a free chair in every meeting. As a silent but permanent reminder of what it's actually about. This form of customer focus also means that the customer is actively involved in the development of services and products. At an early stage, the customer receives prototypes and minimally functional products to give their feedback. Based on this feedback and its needs, agile organizations are constantly developing their offerings. In this way, agile companies gain a high level of clarity at an early stage about what customers really value about their offering and which performance aspects they do not need to pursue further (Michalke, 2021). Management takes on a pioneering role in the agile organization by networking external value creation partners and its own departments with the customer. Production, procedures, and processes are synchronized and regularly checked. Agile supplier management must be organized in such a way that there are units for both the classic purchasing business and for support in exploring complex situations (Kleemann, 2020). Agile working companies work differently than conventional methods. Instead of planning with a high level of detail, many tasks and challenges are tackled at the same time. Agile working does not mean that there is no more planning at all. Agile working means delegating as many tasks as possible to small teams with control and moderation of processes and activities. The cross-departmental teams then work together, including with external partners across company boundaries, on sustainable and quick-to-implement solutions. In order for this to be successful, you need a clearly defined goal and a clear project mandate that need to be achieved. Teams approach solutions in many small steps. They divide a large goal into numerous intermediate tasks (Pfannstiel *et al.*, 2020). This makes it easier to react to changing conditions and constantly adapt. Companies and teams can therefore survive better in an environment that is difficult to predict if they rely on agile methods rather than linear work.

7.5 Advantages of Agile and Lean Organizations

Agile and lean working offer companies clear advantages that are crucial for competitiveness:

- *Quick start*: The project can start quickly because not all details have to be determined in advance.
- *Cost-saving development*: Improvements are implemented at an early stage using short work cycles and constant feedback. This saves time and costs.
- *Optimal coordination with customer needs*: Thanks to constant customer communication, the end product corresponds exactly to the needs of the target group. This not only ensures market success but also prevents the development of unnecessary features.
- *Reliable quality*: Continuous testing of interim results based on clear criteria ensures that quality defects are reliably identified and eliminated.
- *Fast results*: Since a marketable end product is developed right from the start, the time to market is significantly shortened. This is particularly important in the wake of digitalization.
- *Flexibility*: Changing customer needs, new competitive requirements, and technological developments are constantly taken into account.
- *Cost reduction*: Agile project management helps to keep costs under control and even minimize them.
- *Motivated employees*: The agile organization increases the motivation of employees through a cooperative, friendly work culture, shared visions, and by promoting self-determination and creativity.

References

Burns, J. (2018). *Leadership.* New York: Harper Torchbooks.

Fatma, P. (2015). The effect of organizational culture on implementing and sustaining lean processes. *Journal of Manufacturing Technology Management,* 26(Ausgabe 5 S). 725–743.

Gloger, B. (2014). *Wie schätzt man in agilen Projekten-oder wieso Scrum-Projekte erfolgreicher sind.* München: Hanserverlag.

Helmold, M. (2021). *Kaizen, Lean Management und Digitalisierung. Mit den japanischen Konzepten Wettbewerbsvorteile für das Unternehmen erzielen.* Wiesbaden: Springer.

Helmold, M. (2022). *Leadership. Agile, virtuelle und globale Führungskonzepte in Zeiten von neuen Megatrends.* Wiesbaden: Springer.

Helmold, M. & Samara, W. (2019). *Progress in Performance Management. Industry Insights and Case Studies on Principles, Application Tools, and Practice.* Heidelberg: Springer.

Helmold *et al.* (2020). *Corporate Social Responsibility im Internationalen Kontext. Wettbewerbsvorteile durch nachhaltige Wertschöpfung.* Wiesbaden: Springer.

Kleemann, F. (2020). *Agiler Einkauf: Mit Scrum, Design Thinking & Co. die Beschaffung verändern.* Wiesbaden: Springer.

Michalke, A. (2021). *Mitarbeiterführung. Führen als integrative Tätigkeit.* Wiesbaden: Springer.

Mollenhauer, M. & Sommerlatte, T. (2016). *Vertrauensbasierte Führung. Transformationale Führung — vorleben und Inspirieren: Deutsche Bahn.* Wiesbaden: Springer.

Neun, W. (2020). *Digitale Transformation und Agilität in der Praxis. Veränderungsbereitschaft in Unternehmen fördern durch Background-Personality-Management.* Wiesbaden: Springer.

Peters, T. (2015). *Leadership. Traditionelle und moderne Konzepte Mit vielen Beispielen.* Wiesbaden: Springer.

Pendleton, D., Furnham, A., & Cowell, J. (2021). *Leadership. No more Heroes.* Basingstoke: Palgrave Macmillan.

Pfannstiel, M., Siedl, W., & Steinhoff, P. (2020). *Agilität in Unternehmen. Eine praktische Einführung in SAFe und Co.* Wiesbaden: Springer.

Pfeiffer, S. (2021). *Führungskraft: Das große Leadership Buch — Erfolgreiche Mitarbeiterführung durch praxisnahe Methoden und Techniken inkl. Mitarbeitergespräche und Kommunikationstraining.* Eulogia Verlags GmbH Hamburg.

Scherer, K. (2018). Agilität. Und plötzlich sollen die Angestellten selbst denken. *WIWO.* https://www.wiwo.de/my/erfolg/trends/agilitaet-agiles-programm-zum-agilen-arbeiten-bei-adidas/22894528-3.html?ticket=ST-1327538-n9eI-HHKQB7qRSFPakQR0-ap1 (Accessed May 16, 2022).

Chapter 8

Lean Digitization and AI Roadmap

8.1 Scope and Definition

Digitalization and artificial intelligence (AI) are buzzwords in all areas of society and include future-oriented innovations and technologies. The digital transformation and the use of human-like (humanoid) machines are among the most important social and economic developments of our time (Figure 8.1; Schallmo *et al.*, 2023). There are many questions associated with this: What exactly is behind terms such as digital change and the digital transformation of AI?

Digitalization means the use of data and algorithmic systems for new or improved processes, products, and business models. In this context, digitalization includes analogue and digital information for further processing, storage or usage (Sendler, 2020).

AI uses computers and machines to mimic the problem-solving and decision-making abilities of the human mind. Intelligence is understood as the ability to act appropriately and proactively in its environment. This includes the ability to perceive and react to sensory impressions, to absorb information, process it, and store it as knowledge, understand and produce language, to solve problems, and to achieve goals (Rogers, 2023).

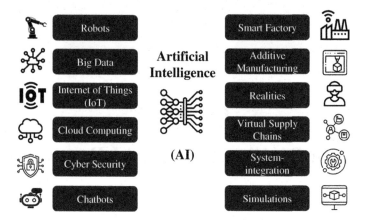

Figure 8.1.   AI elements.

8.2 Digital Maturity Levels

8.2.1 Digital Champions

Digital champions are role models and pioneers for digital transformation within value chains, society, and organizations. Digital champions are companies that offer innovative digital products and services which generate real added value for their employees and customers (Figure 8.2). Digital champions manage to build particularly strong digital customer proximity with their transformation toward digitalization and an online presence on websites, social media profiles, and apps. Figure 8.2 shows the respective maturity levels up to the Digital Champion (Helmold, 2024).

8.2.2 Digital Laggards

Digital laggards are companies that have not yet understood the need for digital transformation. Laggards have a significant gap with the pioneers of digital transformation and can be seen as last or latecomers, lagging far behind the other companies when it comes to digitalization and AI. Laggards often fall by the wayside in competition due to their hesitation (Helmold, 2023).

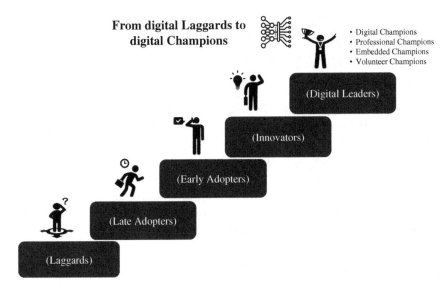

From digital Laggards to digital Champions

- Digital Champions
- Professional Champions
- Embedded Champions
- Volunteer Champions

(Digital Leaders)

(Innovators)

(Early Adopters)

(Late Adopters)

(Laggards)

Figure 8.2.   Digitization maturity levels.

8.2.3 Late Adapters

As compared to the previous year, companies in Germany are primarily recognizing opportunities in the use of AI and seeing more advantages in using the technology. At the same time, the proportion of companies that use AI is only increasing very slowly (Streim, 2023). On the one hand, the late adapters or passive implementers, who have no independent drive to innovate but only react to changing customer requirements, are at the bottom of the pack. Late adapters are hesitant and passive about innovation and the introduction of AI. Many companies wait to introduce the necessary innovations because they are just waiting for them to work. The adoption of digital innovations along the value chain is also a social decision-making process, where if a user talks to a potential user about an innovation that works for the first user, the second user is more likely to adopt that innovation.

8.2.4 Early Adapters

Unlike late adapters, early adapters are not laggards in the introduction and implementation of digitalization measures, but rather companies that partially introduce innovations after carefully weighing up all the advantages and disadvantages.

8.2.5 Innovators

The innovators are pioneers of their time and belong to the group of adapters who are the first to buy a new process or product and thus initiate the spread of innovations. Innovators have a high willingness to take risks. Digital innovations are innovations that result from the innovative use of digital technologies. A central feature of digital innovations is that they are designed in a way, that they multi-purpose or personalized solutions. Innovators also ensure that teams work in an agile, creative and modern way to create this idea-creating environment and organization (Hofert, 2022).

8.2.6 Digital Leaders

The leading drivers and impulse generators in digitalization are digital leaders, professional digital champions, embedded digital champions, and volunteer digital champions. At the forefront of innovation dynamics are the digital leaders as technology leaders who are constantly pushing technological boundaries. They do this with the help of strong basic research and numerous of their own inventions in the area of digitalization along the value chain. Naturally, they have a strong scientific focus on their work. They take care of the strategic implementation of these activities. As digital leaders, they know how important digital integration and transformation are for the success of their business model. If a permanent, systematic change needs to be implemented in the company, they know the way. This is exactly what makes them indispensable as pioneers. Professional digital champions actively drive changes and digitalization activities in companies. They are the driving force in the company. Digital leaders must carefully select professionals and deploy them in a targeted manner so that they can inspire and train their colleagues. Embedded digital champions work in a specific function and specialist departments, such as purchasing, human resources, marketing, sales, and production, and have integrated their commitment as a digital champion into their respective positions. Without it being explicitly stated in their job title, they are champions because they regularly lend a helping hand to their colleagues. If there are topics related to digitalization or tools that require more intensive training, volunteer digital champions are there. They also take unpaid time and support people or organizations wherever necessary (Waller, 2016).

8.3 Digital Roadmap to Become a Champion

Value chains are constantly changing. Due to the growing internationalization of business and the intensification of partnerships, processes are becoming increasingly complex, and demands for efficiency and cooperation are increasing (Forster, 2019). There is also a growing need to evaluate information and data and to create the greatest possible transparency about markets and customers. And for some time now, companies have been increasingly looking for new business areas and potential. This requires digital strategies and innovative IT solutions to achieve the goals of the business (Engelen & Schneider, 2021). This increases the pressure on management to develop a digital roadmap for their company (Machado & Davim, 2023). The aim of a business-oriented digitalization roadmap is to optimally align all value-creation activities of internal and external stakeholders with the company's goals and business processes (Figure 8.3). In the first step, project managers and project team(s) are selected. External consultants and experts can also be involved for this purpose. In the second phase, an analysis of the current situation takes place. In the third step, the roadmap is defined. In this phase, measurable, realistic, and implementable goals are defined. In the penultimate and fourth step, the strategy is implemented, and then in the fifth step, it is checked for successful and sustainable implementation.

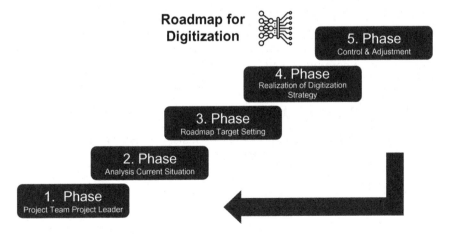

Figure 8.3.   Digital roadmap.

8.4 Dealing with Resistance in Digitalization and AI

8.4.1 First Strategy: Analyze Internal Stakeholder Groups

Every digitalization project has different interest groups and stakeholders (Helmold *et al.*, 2022). Stakeholders (synonymously, project participants) are people, groups of people, associations, or organizations who are affected by the project or its results or who can influence the project. Internal stakeholders in particular are often forgotten or underestimated; the focus is usually on the customer or the public. Experienced project managers therefore always analyze which groups in the company are affected by their project and then try to gain clarity about the individual groups:

- Which stakeholders are affected?
- How are the stakeholders affected?
- What are the interests of the individual groups?
- What behavior can be expected from the individual groups?
- What pattern of action are they likely to follow?

8.4.2 Second Strategy: Explain the Background to the Transformation and the Digitalization Project

Projects bring changes and changes offer opportunities. But these opportunities alone cannot convince all stakeholders of a project. Real orientation provides answers to the question of why a change project is being carried out. Only then will it be explained how the project should run and what exactly will change. In this context, it is necessary to explain the reasons for the project. Targets, time, financial, personnel, or other conditions, demarcations from other projects, and project-specific organization should also be communicated.

8.4.3 Third Strategy: Deal with Opponents Correctly

Almost every project has opponents who resist change. The only question is how the project manager optimally behaves toward these opponents or resistance fighters and how he approaches them. Resistance can be carried out openly or covertly. Fear or insecurity is often the reason for resistance. Fear, for example, about cherished habits and familiar surroundings. Or the fear of suffering disadvantages in the project. The biggest mistake when dealing with opponents is withholding and concealing bad news (Helmold *et al.*, 2022) because many people have a good gut feeling about the truth. They anticipate the consequences of projects. And the feeling that something is being kept secret increases mistrust and resistance. That's why project professionals provide early and open information about the dark side of the project.

8.4.4 Fourth Strategy: Preserve and Develop What is Good

Many stakeholders are not fundamentally opposed to the innovations of a project. However, they want proven and good things to be retained. Project managers should not dismiss this interest of the "keepers." On the one hand, project managers tend to do "tabula rasa" and unnecessarily delete tried-and-tested things. On the other hand, it helps stakeholders if, despite all the innovations, they continue to find what they know and value. For example, you can incorporate good things that you find into the change concept — and even improve them even further by cleverly combining them with the new. Experienced project managers provide clear and emphatic information about what will not be changed in the project and will remain valuable (Helmold *et al.*, 2022).

8.4.5 Fifth Strategy: Awareness of the Valley of Tears

A problem, especially with organizational change and change management projects, is the negative aspects that arise, as innovations such as converted processes and procedures or organizational changes only really work correctly after a learning curve. These negative aspects often lead to a "valley of tears" among employees. Changes disrupt a system that has often previously functioned well; time usually passes until the changed system has stabilized again, but then at a higher level than before. Project managers call this period of time the "valley of tears," i.e., the weeks and months in which the promised benefits do not (yet) materialize. The stakeholders must therefore be made aware of the "valley of tears" early on and be made to understand that worthwhile change also requires patience. Otherwise, the project manager forfeits acceptance. The "valley of tears," also known as the change curve, is the turning point in the change process. Those affected realize that there is no way around saying goodbye to old habits and adapting to the new reality.

8.4.6 Sixth Strategy: Provide Individual Information

As important as the background and procedures in a project are, the stakeholders want to know above all how the project will affect their everyday work, specifically and individually. Project managers therefore develop an individual information strategy for each stakeholder group. A four-step approach is recommended for individual information (Helmold *et al.*, 2022). First, explain to each group what is valuable in their area and what will be preserved. Second, what will each group lose? What will the project change for the individual, and what will she have to do without in the future? Third, what does the individual gain from the change? And fourth, what opportunities and advantages arise for the whole, for example, for the company or the client? Some project managers constantly present the overall benefits to the company. Strategically, it is cleverer to talk about it at the end and to first understand and meet the needs of the individual (Helmold *et al.*, 2022; Helmold, 2022).

References

Engelen, A. & Schneider, O. (2021). *Die Strategien Digitaler Champions.* Springer Wiesbaden.

Forster, N. (2019). *Hidden Digital Champions: Wie sich KMUs und das Handwerk für die Zukunft rüsten (Fit for Future).* Wiesbaden: Springer.

Helmold, M. (2022). *Leadership. Agile, virtuelle und globale Führungskonzepte in Zeiten von neuen Megatrends.* Wiesbaden: Springer.

Helmold, M. (2023). *Wettbewerbsvorteile entlang der Supply Chain sichern. Best-Practice-Beispiele in Beschaffung, Produktion, Marketing und anderen Funktionen der betriebswirtschaftlichen Wertschöpfungskette.* Wiesbaden: Springer.

Helmold, M. (2024). *Erfolgreiche Transformation zum digitalen Champion: Wettbewerbsvorteile durch Digitalisierung und Künstliche Intelligenz.* Wiesbaden: Springer.

Helmold, M. *et al.* (2022). *Lean Management, Kaizen, Kata and Keiretsu. Best-Practice Examples and Industry Insights from Japanese Concepts.* Wiesbaden: Springer.

Hofert, S. (2022). *Agiler führen. Einfache Maßnahmen für bessere Teamarbeit, mehr Leistung und höhere Kreativität.* Wiesbaden: Springer.

Machado, C. & Davim, J. P. (2023). *Management for Digital Transformation.* Cham: Springer.

Rogers, D. L. (2023). *The Digital Transformation Roadmap. Rebuild Your Organization for Continuous Change.* Columbia: eBook Columiba Press.

Schallmo, D. R. *et al.* (2023). *Digitalisierung: Fallstudien, Tools und Erkenntnisse für das digitale Zeitalter (Schwerpunkt Business Model Innovation).* Wiesbaden: Springer.

Sendler, U. (2020). *KI-Kompass für EntscheiderKünstliche Intelligenz in der Industrie: Strategien — Potenziale — Use Cases.* München: Hanser Verlag.

Waller, S. (2016). *The Digital Champion: Connecting the Dots between People, Work and Technology.* Online Book.

Chapter 9

Lean Strategy Recommendations — Creating a Competitive Advantage

9.1 Need for Cultural Change Toward Kaizen

Successful organizations do not prosper by devoting a ruthless approach to chip away at costs, relentlessly reducing all decision-making to a reduction in headcount. The purpose of lean culture change is to secure the future of the entity by uniting its people to deliver the voice of the customer (Helmold, 2024; Helmold *et al.*, 2022). In the not-for-profit sector, lean culture change drives organizational success by adding value to existing consumers and winning loyalty (Pascual-Rueno *et al.*, 2019). It is about developing resilient service provision, developing core staff competencies, and attracting and retaining the best people. The lean culture reverses the polarity of the organization, shifting from a fire-fighting mode to a planning mode, where prevention of problems rather than reacting after the event of failure becomes the norm. Healthy, positive, and organizational cultures are characterized by a long-term continuity perspective, with a focus on tactics to resolve immediate short-term problems (Machado & Davim, 2023). The dominant culture should support and reward cross-organizational working (Engelen & Schneider, 2021; Helmold, 2023). Implementing lean thinking is a cultural change that requires leadership because in the end, it's all about people. Here are 10 guidelines your leader can adopt right now to change the culture.

9.1.1 Challenge People to Think

If you are not thinking, you're not learning new things. If you're not learning, you're not growing — and over time becoming irrelevant in your work. The most successful leaders understand their colleagues' mindsets, capabilities, and areas for improvement. They use this knowledge/insight to challenge their teams to think and stretch them to reach for more.

9.1.2 Lead by Example

Leading by example sounds easy, but few leaders are consistent with this one. Successful leaders practice what they preach and are mindful of their actions (Dathe *et al.*, 2022; Helmold, 2022). They know everyone is watching them and therefore are incredibly intuitive about detecting those who are observing their every move, waiting to detect a performance shortfall.

9.1.3 Take Lots of Leaps of Faith

Making a change requires a leap of faith. Taking that leap of faith is risky, and people will only take active steps toward the unknown if they genuinely believe — and perhaps, more importantly, feel — that the risks of standing still are greater than those of moving forward in a new direction. Making a change takes lots of leaps of faith.

9.1.4 Create an Environment Where It is Acceptable to Fail

Failure should be encouraged! That's right. If you don't try, you can't grow; and if growth is what you seek, failing is inevitable. There must be encouragement to try, and it's ok if you try and it doesn't work. An environment where you can't fail creates fear.

9.1.5 Eliminate Concrete Heads

"Concrete Heads" is the Japanese term for someone who does not accept that the organization must be focused on the elimination of waste. People feel threatened by the changes brought about by lean. As waste and bureaucracy are eliminated, some will find that little of what they have been doing is adding value. The anxiety they feel is normal and expected. To counteract this, it is critical that people are shown how the concept of work needs to change.

9.1.6 Be a Great Teacher

Successful leaders take the time to mentor their colleagues and make the investment to sponsor those who have proven they are able and eager to advance. They never stop teaching because they are so self-motivated to learn themselves.

9.1.7 Show Respect to Everyone

Everyone desires respect. Everyone. Regardless of your position or power, ensure you show everyone respect. Everyone wants to be treated fairly.

9.1.8 Motivate Your Followers

Transformational leaders provide inspirational motivation to encourage their followers to get into action. Of course, being inspirational isn't always easy. Some ideas for leadership inspiration include being genuinely passionate about ideas or goals, helping followers feel included in the process, and offering recognition, praise, and rewards for people's accomplishments.

9.1.9 Develop a True Team Environment

Create an environment where working as a team is valued and encouraged and where individuals work together to solve problems and help move the organization forward. Individuals who will challenge each other and support each other make teams more successful.

9.1.10 Encourage People to Make Contributions

Let the members of your team know that you welcome their ideas. Leaders who encourage involvement from group members have been shown to bring about greater commitment, more creative problem-solving, and improved productivity. Constant change is a business reality and organizations must continually adapt to their environments to stay competitive or risk losing relevance and becoming obsolete. For each change, leaders must define it, create a vision of the post-change world, and mobilize their teams to make it. Fundamentally, a change of culture occurs when people start behaving differently as a result of a change in the climate of the organization. There are many different models of how an organizational culture is shaped by the prevailing climate and how it can be assessed. Leaders who protect the *status quo* through control must surrender to change in order to secure the future of their organization. Don't be the leader who rewards herd mentality and "me too" thinking. Don't be the leader who encourages people not to fail or not to take risks. Be the leader who both models and gives permission to do the exact opposite of the aforementioned — be a leader who leads. The culture of an organization is learned over time. It can be taught to new employees through formal training programs but is more generally absorbed through stories, myths, rituals, and shared behaviors within teams. Organizational culture will have a positive or negative impact on everything you try to do whether you want it to or not.

9.2 Transformation and Adaptability of Strategies and Tools

Fierce competition, globalization, and the recent COVID-19 pandemic are leading to significant challenges for enterprises and organizations. As, a consequence, it is necessary to have appropriate countermeasures, corrective actions, and tools to overcome crises. On the contrary, those enterprises and organizations that don't have the right strategies and management tools will not survive (Lauer, 2019).

In this context, challenges in management are characterized by high complexity, unpredictability, and uncertainty. Managers must therefore have effective, quick, and pragmatic concepts for structuring and resolving problems. The classic management literature is often too theoretical or too detailed for this. A short, concise introduction to a tool is sufficient to give modern managers the food they need for thought (Kieviet, 2019). This is where this book starts. It clearly presents effective strategies and management tools, as well as the most important concepts. It describes possible applications and makes it easier to interpret the results. The book offers the manager pragmatic and effective help to gain transparency about existing concepts and tools, to find the right tool for the respective situation, and ultimately to use it effectively to obtain a long-term sustainable competitive advantage (Helmold & Samara, 2019).

As a rule, the need for corporate transformation and change is associated with changing market conditions or economic, technological, or social changes, as a result of which the company suffers from a decline in sales, rising operating costs, and decreasing customer relevance or customer loyalty. Often, it is the big trends and developments, such as digitization or the trend toward sustainability. The example of the publishing house or the music industry makes it clear how existentially threatening such changes can be to a company. These examples also make it clear that business transformation is not just a question of the need for change but also a question of the ability to change. While economic, technical, and social changes represent an existential threat for one company, they are a guarantee of success for adaptable and agile companies and offer a wide range of business opportunities. E-commerce, cloud computing, and the green economy are intended to illustrate this symbolically at this point. Entire new industries and business fields are emerging here that would never have emerged without technological change and social and political pressure. Even if some companies suffer from constant, increasingly accelerating change, as a rule every change brings improvement with it — improvement for consumers, the environment, and society. So, change means evolution. The art of making use of change for the benefit one's own company therefore lies in continuously helping consumers, the environment, and society to improve their quality of life. This creates new ideas, new business areas, new revenue models, new industries and ultimately lasting success, supported by more sales and more emotional and loyal customer relationships. With this in mind, whenever possible, contribute to improving the quality of life of your customers, the environment, and society — and always remain adaptable.

9.3 Lean Transformation Strategies

Lean management transformation describes an optimizing change in business activity and/or working methods emanating from the company management, which can only include partial areas as well as the company as a whole. The need for lean management transformation is usually based on external factors. These can be, for example, tougher competitive conditions, fundamental changes in the law, or general social and technological change, such as that brought about by digitization or the trend toward sustainability.

The aim of the business transformation is always to strategically secure business operations over the long term, thus creating the basis for lasting and sustainable success (Klasen, 2019). It is not an approach to the realization of short-term, one-off effects. It does not concentrate on individual fields of action, but rather captures the change process as a whole. In addition, the focus of the business transformation is not on individual persons or groups of people, but rather integrates the entire economic, social and societal environment of the company. Lean management transformation strategies aim:

- To secure and increase sales sustainably.
- To reduce operating and other costs.
- To intensify customer satisfaction and customer loyalty.

The term was first used by the management consultancy Gemini Consulting in the early 1990s. Their consultants Francis J. Guillart and James N. Kelly published a book in which they processed the experiences from their everyday work and at the same time detailed lean management transformation as a formula for the long-term success of companies in a constantly changing market economy.

The basis of their considerations was the view of companies as living organisms. This should enable them to react flexibly to fluctuations and changes in the market in order to insist on it permanently, economically, and profitably. Lean management transformation projects are usually very complex and lengthy. There are several reasons for this. On the one hand, lean management transformation projects tend to be used in larger companies. This is less due to the size of the company itself or the financial possibilities, but rather because the management of a stock corporation or group has to answer to shareholders, partners, investors, and supervisory boards, whose own financial future depends heavily on the success of the company. Owner-run companies, on the other hand, often suffer from sticking to the tried-and-tested rituals for too long. The emotional relationship with the past is significantly higher and often proves to be the greatest brake on change in management and the workforce.

In addition, many lean management transformation projects are initiated far too late by those responsible. Often, the course of the crisis or the process of economic, technological, or social change is already very far advanced. And companies can often no longer make up for the lost time. One industry that has had to experience this in a painful way is the photography industry. Former industry giants such as Kodak, Agfa, or Pollaroid, who, despite great efforts, were never able to build on the great successes of bygone days. The principle of hope seems omnipresent, and statements such as "this trend will definitely pass" or "we'll get in when the others have made their mistakes" are not uncommon. This may sound ironic to some. It has always been one of the outstanding virtues of German entrepreneurship to develop and occupy markets through innovations, inventiveness, and engineering skills.

Since lean management transformation strategies are essential for the future and sustainable success of the company, these projects are always the responsibility of the management. In addition to a large number of internal and external experts and industry and company experts, lean management transformation projects are usually accompanied by a specialized management consultancy that guides the company through the four relevant project sections of lean management transformation in a targeted and efficient manner (Womack *et al.*, 1990). The reframing goal setting and goal definitions are:

- Restructuring — structural reorientation of the company.
- Revitalizing — product development and innovation.
- Renewing — consistent internal and external implementation.

References

Dathe, T. *et al.* (2022). *Corporate Social Responsibility (CSR), Sustainability and Environmental Social Governance (ESG). Approaches to Ethical Management.* Cham: Springer.

Engelen, A. & Schneider, O. (2021). *Die Strategien digitaler Champions.* Wiesbaden: Springer.

Helmold, M. (2022). *Leadership. Agile, virtuelle und globale Führungskonzepte in Zeiten von neuen Megatrends.* Wiesbaden: Springer.

Helmold, M. (2023). *Wettbewerbsvorteile entlang der Supply Chain sichern. Best-Practice-Beispiele in Beschaffung, Produktion, Marketing und anderen Funktionen der betriebswirtschaftlichen Wertschöpfungskette.* Wiesbaden: Springer.

Helmold, M. (2024). *Erfolgreiche Transformation zum digitalen Champion: Wettbewerbsvorteile durch Digitalisierung und Künstliche Intelligenz.* Wiesbaden: Springer.

Helmold, M. *et al.* (2022). *Lean Management, Kaizen, Kata and Keiretsu. Best-Practice Examples and Industry Insights from Japanese Concepts.* Wiesbaden: Springer.

Lauer, T. (2019). *Change Management. Der Weg zum Ziel.* Wiesbaden: Springer.

Machado, C. & Davim, J. P. (2023). *Management for Digital Transformation.* Cham: Springer.

Womack, J. *et al.* (1990). *The Machine That Changed the World: The Story of Lean Production, Toyota's Secret Weapon in the Global Car Wars That Is Now Revolutionizing World Industry.* New York: Free Press.

Chapter 10

Lean Production System

10.1 Scope and Definition

The lean production system, or Toyota production system (TPS), contains a set of principles which help enterprises to successfully transform into lean organizations (Helmold *et al.*, 2022). As an initial step, it is important to apply the 5S concept (Figure 10.1), followed by the lean production principles (Figure 10.2).

10.2 5S Concept

The 5S concept is the name of a workplace organization method that uses a list of five Japanese words: *seiri* (整理), *seiton* (整頓), *seiso* (清掃), *seiketsu* (清潔), and *shitsuke* (躾). Transliterated into Roman script, they all start with the letter "S." 5S is used to stabilize, maintain, and improve the safest and best working environment, thus supporting sustainable quality, cost/finance and delivery (QCD)-plus alpha. 5S is a systematic and structured workplace optimization, originally developed and used by Toyota. The objective is the identification and elimination of waste. In simple terms, the 5S methodology helps a workplace remove items that are no longer needed (sort), organize the items to optimize efficiency and flow (straighten), clean the area in order to more easily identify problems (shine), implement color coding and labels to stay consistent with other areas (standardize), and develop behaviors that keep the workplace organized over the long term (sustain) (Pinto *et al.*, 2018).

Figure 10.1.　5S concept.

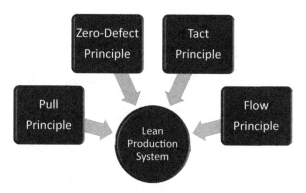

Figure 10.2.　Lean production system.

These five words can be translated as "Sort," "Set in order," "Shine," "Standardize," and "Sustain." The 5S-methodology describes how to organize a work space for efficiency and effectiveness by identifying and storing the items used, maintaining the area and items, and sustaining the new order (Niemann *et al.*, 2021). The decision-making process usually comes from a dialogue about standardization, which builds understanding among employees of how they should do the work. In some quarters, 5S has become 6S, the sixth element being safety or self-discipline. The first element in the 5S concept is sorting (*seiri*). In this step, it is important to distinguish between necessary and unnecessary things. Things in this context are materials, components, tools, gauges, information, and people. Unnecessary things must disappear. Removing these items which are not used in the working area may take a reasonable amount of time. Classification of all equipment and materials by frequency will help to decide if these items can be removed or not. The second step is setting in order (*seiton*). This is the practice of orderly storage, so the right item can be picked efficiently at the right time and is easy to access for the operators. Identification and allocation of materials, information, tools, and necessary things at fixed and visualized locations is important in this step. In the next and third step (*seiso*), it is mandatory to create a clean worksite without garbage, dirt, and dust, so problems can be more easily identified (leaks, spills, excess, damage, etc). In the fourth step, (*seiketsu*) standards for a neat, clean workplace and operations will be set up through visual management. In the fifth and last stage (*shitsuke*), it is important to create the environment, patterns, management style, and behavior to ensure that established standards are executed over the long term, making the workplace organization the key to managing the process for success (Helmold & Terry, 2021). The advantages of the 5S System are as follows:

- Creation of a transparent layout;
- Visualization of processes;
- Transparency of hidden and obvious waste;
- Elimination of unnecessary activities;
- Improvement of efficiency;
- Focus on safety measures;
- Increase employee motivation through simplification of the work environment;
- Ensuring that all materials are instantly available;
- Ensuring that tools (screwdriver, devices) are readily reachable;
- Ensuring that required (work procedures, work sequence, etc.) information is instantly available;
- Reduction of waste.

10.3 Lean Production Principles

10.3.1 Pull System

The pull system principle is one of the lean production principles and is used to reduce waste in the production process. In this type of system, components used in the manufacturing process are only replaced once they have been consumed, so companies only make enough products to meet customer demand. The opposite principle is the push system, in which as many products as possible are generated to be sold via marketing activities. The principles aim to avoid overproduction and stockpiling, thereby saving working capital, by letting demand dictate the rate at which goods or services are delivered (Aberdeen Group, 2005). In this way, the customer, or the next step in the chain, "pulls" value through the process. The pull system should be ideally installed across the value chain and value chain partners (Bothard *et al.*, 2009).

10.3.2 Zero Defect System

Zero defects are the principle of thinking and doing that reinforces the notion that defects are not acceptable and that everyone should do things right the first time. The major goal is to eliminate errors and defects across the value chain in all production or service steps. Zero defects theory should not be confused with perfectionism, however. The philosophy of zero defects doesn't mean that people never make mistakes. Instead, it means that companies shouldn't be working with the idea that not giving one's best is the norm. Substandard and unclear targets where the final product can just be fixed later should not be expected; instead, management should work to create things right on the first try with clear targets to produce their ideal outcomes.

10.3.3 Tact System

Tact refers to the rhythm at which goods or services are produced to meet customer demand. With a consistent, continuous rhythm providing a heartbeat for your production processes, it is far easier to regulate, responding flexibly and effortlessly as demand rises or falls. Tact time is defined as the average time available (time available minus breaks, maintenance, or set-up) divided by the customer-requested quantity, as shown in Figure 10.3.

Figure 10.3.   Tact system.

10.3.4 Flow System

Flow is how work progresses through a system. When a system is working well, or having a "good" flow, it tends to move steadily and predictably, whereas, a "bad" flow means the work starts and stops (Sinha & Matharu, 2019). Every time there is a breakdown in the flow, the chances of accumulating waste increase. One goal is to strive for a consistent flow that generates more reliable delivery and greater value to customers, teams, and stakeholders (Ohno, 1990). There are seven types of flow:

(1) The flow of raw materials;
(2) The flow of work-in-process;
(3) The flow of finished goods;
(4) The flow of operators;
(5) The flow of machines;
(6) The flow of information;
(7) The flow of engineering.

Companies that successfully integrate the principles of lean production understand that when each of these seven types of flows is working in harmony, they are increasing their odds of producing finished goods and services that require little to no corrective action (Liker, 2004). Keeping the production running in such a smooth fashion also helps to ensure that a company is creating efficiencies (Figure 10.2).

10.4 Andon

Andon (Japanese: アンドン or あんどん or 行灯) is a lean manufacturing tool referring to a system to notify management, maintenance, and other workers of a quality or process problem. The centerpiece is a device incorporating signal lights to indicate which workstation has the problem. The alert can be activated manually by a worker using a pull cord or button, or it may be activated automatically by the production equipment itself. The system may include a means to stop production so the issue can be corrected. Some modern alert systems incorporate audio alarms, text, or other displays. An Andon system is one of the principal elements of the Jidoka method pioneered by Toyota as part of the TPS and therefore now part of the lean concept. It gives the worker the ability, and moreover the empowerment, to stop production when a defect is found and immediately calls for assistance. Common reasons for manual activation of the Andon are part shortage, defect created or found, tool malfunction, or the existence of a safety problem. Work is stopped until a solution has been found. The alerts may be logged into a database so that they can be studied as part of a continuous-improvement program. The system typically indicates where the alert was generated, and may also provide a description of the trouble. Modern Andon systems can include text, graphics, or audio elements. Audio alerts may be done with coded tones, music with different tunes corresponding to the various alerts, or pre-recorded verbal messages. Usage of the word originated within Japanese manufacturing companies, and in English, it is a loanword from a Japanese word for a paper lantern (Imai, 1986). Figure 10.4 shows an Andon example at Alstom in China. The red light means the disruption of production in the respective production operation.

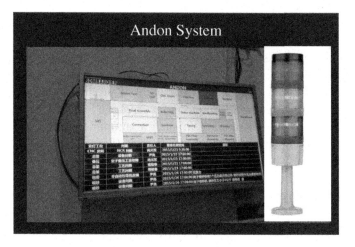

Figure 10.4.   Andon system.

10.5 Shadow Boards

Shadow boards are specific boards for parts, tools, and equipment in operations, manufacturing, or service areas to reduce waste and waiting time (Helmold *et al.*, 2022). The aim of the shadow board is to achieve an organized workplace where tools, supplies, and equipment are stored in appropriate locations close to the work area or workstations. It provides the basis for standardization in the workplace. They are a simple and inexpensive tool that provides tangible efficiencies and cost savings as well as intangible benefits. Figure 10.5 shows a shadow board for screws in Mitsubishi Japan. The appropriate storage, allocation, and preparation of screws avoid waiting time and the possibility of errors. The advantages of using shadow boards include avoiding waste, such as time spent looking for the appropriate tool or even having to buy a new one, and wasted time in looking for supplies and interchanging tools between tasks. Shadow boards also provide the ability to quickly gauge the location of tools and equipment or if they are missing. Shadow boards are used in the sort and set in order stages of the implementation and operation of a 5S system in

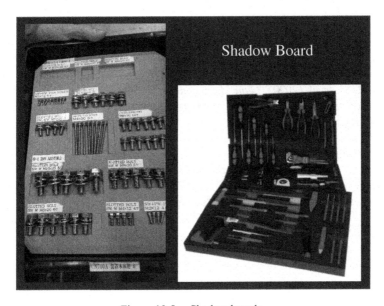

Figure 10.5.   Shadow boards.

a workplace and *kaizen* initiatives. Shadow boards can be different sizes and located in many different areas of a process or plant. The key is that they are appropriately located and hold all the necessary tools for the area or workstation.

10.6 Poka Yoke

Poka Yoke (ポ カ ヨ ケ) is a Japanese term that means "mistake-proofing." A *poka-yoke* is any mechanism in a lean concept that helps an equipment operator avoid (*yokeru*) mistakes (*poka*). Its purpose is to eliminate product defects by preventing, correcting, or drawing attention to human or other errors as they occur. The concept was formalized and the term adopted by Shigeo Shingo as part of the TPS. It was originally described as *baka-yoke*, but as this means "fool-proofing" (or "idiot proofing"), the name was changed to the milder *poka-yoke* (Bertagnolli, 2020).

References

Bertagnolli, F. (2020). *Lean Management. 2. Auflage. Einführung und Vertiefung in die japanische Management-Philosophie*. Wiesbaden: Springer.

Aberdeen Group. (2005). *Assuring Supply and Mitigating Risks in an Uncertain Economy. Supply Risk Management Benchmark*. Boston.

Helmold, M. & Terry, B. (2021). *Operations and Supply Management 4.0: Industry Insights, Case Studies and Best Practices*. Cham: Springer.

Helmold, M. *et al.* (2022). *Lean Management, Kaizen, Kata and Keiretsu: Best-Practice Examples and Industry Insights from Japanese Concepts*. Cham: Springer.

Imai, M. (1986). *Kaizen. Der Schlüssel zum Erfolg der Japaner im Wettbewerb*. Frankfurt: Ullstein.

Liker, J. K. (2004). *The Toyota Way*. Madison: Mc Graw-Hill.

Niemann, J., Reich, B., & Stöhr, C. (2021). *Lean Six Sigma. Methoden zur Produktionsoptimierung*. Cham: Springer.

Ohno, T. (1990). *Toyota Production System. Beyond large Scale Production*. New York: Productivity Press.

Sinha, N. & Matharu, M. (2019). A comprehensive insight into lean management: Literature review and trends. *Journal of Industrial Engineering and Management*, 12(2).

Chapter 11

Lean Concept of Gemba, Genbutsu, and Genjitsu

11.1 Scope and Definition

Gemba, Genchi, and Genbutsu (Figure 11.1) is a lean methodology which focuses on seeing and understanding the actual work of a process or system with one's own eyes in order to make informed decisions. It involves going to the source of an issue in order to find out what is really happening and how it can be improved. It encourages employees to take ownership of their problems and to find solutions that will benefit both the organization and its customers.

11.2 Gemba

Gemba (現場) is a Japanese word that means "the true place" or "the place where value is created." In lean management, Gemba refers to the place where your process occurs, e.g., a factory hall, a customer service center, or a warehouse. When you visit Gemba, you can observe the actual workflow, human-machine interactions, sources of waste and variation, and the voice of the customer. Gemba helps you gain a deeper and more accurate understanding of your process and its performance.

To make Gemba effective, you should plan your visit, prepare questions, and respect the people who work there. First, define the purpose and scope of your visit and include process owners and stakeholders. Ask open-ended questions to explore the process, and observe and listen more

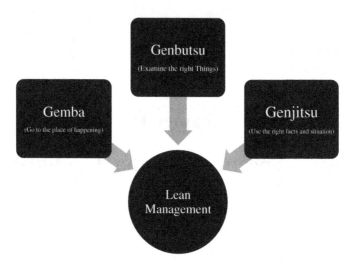

Figure 11.1.   Gemba–Genbutsu–Genchi.

than you speak. Don't forget to take notes, photos, or videos to document your results. Additionally, thank the staff working at Gemba for their time and cooperation, share your observations, and ask for their feedback and suggestions. In this context, it is a significant task of mangers to integrate the 3G approach into the company's values and culture (Ohno, 1990). Successful enterprises have a lean management and shopfloor management focus, as the shopfloor is producing the products or services for their customers (Brunner, 2023). In lean management companies, this approach is called Gemba-Kaizen (Imai, 2012). A central success factor is here the empowerment of shopfloor employees with their contribution towards improvements (Liker, 2004).

11.3 Genbutsu

Genbutsu (現物) means "the actual thing" or "the actual product." In problem-solving, this principle encourages you to examine the actual components, materials, or data related to the issue. By closely inspecting the Gembutsu, you can identify defects or inefficiencies and gain a clear picture of what needs to be addressed (Bertagnolli, 2020).

11.4 Genjitsu

Genjitsu (現実) translates to "the current reality" or "the truth." It requires acknowledging the reality of the situation without making assumptions or judgments (Helmold & Terry, 2021). This principle encourages objectivity and data-driven decision-making. It involves collecting and analyzing data to understand the problem's scope and impact accurately (Helmold *et al.*, 2022).

11.5 From 3G to 5G

11.5.1 Scope and Definition

The 5G method in Figure 11.2 is the ideal tool to concentrate on the relevant lean elements. It is derived from the five Japanese words Gemba, Gebutsu, Genjitsu, Genri and Gensoku. By delving into each of the five categories, it is possible to concentrate on the important issues in the shopfloor. This approach aims at concentrating on important aspects in the production or service area, emphasizes the standardization and focuses on customer satisfaction.

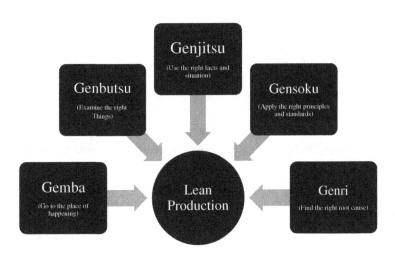

Figure 11.2. 5G model.

11.5.2 Genri

Genri (原理) means "the theory" or "the principle." This step involves identifying the root causes of the problem. Instead of merely addressing symptoms, you dig deeper to find the underlying principles that have led to the issue. Understanding the fundamental principles allows for sustainable and effective solutions (Niemann *et al.*, 2021).

11.5.3 Gensoku

Gensoku (原則), the final "G," means "the standard" or "the rule." This principle focuses on creating and implementing standardized processes and guidelines to prevent the recurrence of the problem. It involves setting new standards based on the lessons learned during the problem-solving process (Helmold, 2020; Helmold *et al.*, 2022).

The 5G Japanese method is not limited to a specific industry or sector. It can be applied in manufacturing, healthcare, software development, or any field where problems arise. By following these five principles, you can develop a comprehensive problem-solving approach that fosters continuous improvement and innovation within your organization. To implement the 5G method effectively, consider the following steps:

(1) *Identify the problem*: Clearly define the problem and its impact on your organization.
(2) *Go to the Gemba*: Visit the actual location where the problem exists.
(3) *Examine the Gembutsu*: Analyze the physical components or data associated with the problem.
(4) *Understand the Genjitsu*: Collect and analyze data to gain a realistic view of the situation.
(5) *Discover the Genri*: Identify the root causes of the problem.
(6) *Implement the Gensoku*: Establish standardized processes to prevent future occurrences.

References

Bertagnolli, F. (2020). *Lean Management. 2. Auflage. Einführung und Vertiefung in die japanische Management-Philosophie.* Wiesbaden: Springer.

Brunner, F. J. (2023). *Japanische Erfolgskonzepte: KAIZEN, KVP, Lean Production Management, Total Productive Maintenance, Shopfloor*

Management, Toyota Production System, GD³ – Lean Development. München: Hanserverlag.

Helmold, M. (2020). *Lean Management and Kaizen. Fundamentals from Cases and Examples in Operations and Supply Chain Management.* Cham: Springer.

Helmold, M. & Terry, B. (2021). *Operations and Supply Management 4.0: Industry Insights, Case Studies and Best Practices.* Cham: Springer.

Helmold, M. *et al.* (2022). *Lean Management, Kaizen, Kata and Keiretsu. Best-Practice Examples and Industry Insights from Japanese Concepts.* Cham: Springer.

Imai, M. (2012). *Gemba Kaizen: A Commonsense Approach to a Continuous Improvement Strategy,* Second Edition (Informatica), New York: McGraw Hill.

Liker, J. K. (2004). *The Toyota Way.* Madison: Mc Graw-Hill.

Niemann, J., Reich, B., & Stöhr, C. (2021). *Lean Six Sigma. Methoden zur Produktionsoptimierung.* Cham: Springer.

Ohno, T. (1990). *Toyota Production System. Beyond large Scale Production.* New York: Productivity Press.

Chapter 12

Lean Audits, Audits, and Quality Management Systems (QMS)

12.1 Lean Audits

12.1.1 Scope and Definition

Lean management principles are based on the idea that a company is at its best when it minimizes physical waste and maximizes its available resources. It emphasizes the constant improvement of processes and activities in the value chain toward customer satisfaction. Implementing lean manufacturing policies and procedures in your company is a great way to increase company productivity and operational efficiency, as well as save money. To ensure that lean practices are correctly used throughout companies and their value chain activities, lean audits must be carried out (Paterson, 2015). Several methods can be applied to conduct an informative or formal lean audit. The purpose of a lean management audit is the exploration of whether the processes and activities are fairly presented in conformity with appropriate lean principles (Helmold *et al.*, 2023).

12.1.2 Phases of Lean Audits

Lean management audits can be broken up into three major steps (and five phases), as shown in Figure 12.1:

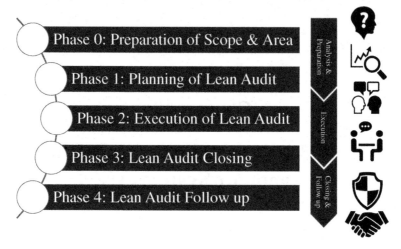

Figure 12.1.   Audit phases and steps.

(1) Analysis and Preparation
(2) Execution
(3) Closing & Follow up

Phase zero is the preparation and analysis phase. Auditors have to define and narrow down the scope of the audit. The scope can focus on a specific or generic process, a department, an activity, or an element of the value chain. A lean audit should be conducted to determine if the business is properly implementing and upholding lean methodologies. This is achieved by analyzing the company's current lean processes with the goal of recognizing opportunities to improve procedures and reduce company costs. This phase is also an opportunity to understand the audited area better and to obtain pre-information such as work or process descriptions, documents, company information, procedures, and relevant tools.

Phase one targets the actual and more specific planning of the lean audit. Company audits should ideally be carried out by a third party or specialist audit team with no vested interest in the outcome of the audit. If the audit is performed internally, qualified lean experts and facilitators should be used. Based on preparation, lean audits can be carried out by segmenting and breaking down manufacturing areas or processes. The scope should be communicated to the involved departments, sections, and employees, which will be part of the audit. A lean audit project plan will help to address the right scope, elements, and stakeholders. With transparent objectives, managers, stakeholders, and employees will be committed to the audit's success. Figure 12.2 (Work Aid 01) shows an example of a lean audit plan.

Project Scope and Plan – Lean Audit						
Value Chain Element, Area, Process, Supplier, Department Description						
Lean Auditor (Expert)r						
Stakeholders, Audit Participants						
Scope: (Detailed Description)						
Objectives: (Expert) S: Specific M: Measurable A: Attainable R: Relevant T: Timely						
Organization & Experts						
Management	Name	Department	Function	Phone	Capacity Engagement	
Audit Team	Name	Department	Function	Phone	Capacity Engagement	
Lean Audit Project Plan (Detailed Description in Attachments)						
Preparation, Kick Off	**Analysis & Planning**	**Execution Review**	**Execution Review**	**Closing & Follow up**		
(Date)	((Date)	(Date)	(Date)	(Date)		
Lean Auditor Expert (Date)		**Management** (Date)		**Stakeholder** (Date)		

Figure 12.2. Lean audit project plan.

Phase two is the actual execution of the lean audit, including the assessment criteria, scoring model, and layout of questions. Lean audit criteria are based on a standard set of tools, questions, and norms but should be customized and tailored to each business, using key lean components like inventory, supply chain management, and procedure standardization. The outcome of the audit can be negatively impacted by the lack of a point system, differing auditing methods, and selective employee interviews.

Phase three is the official closing of the audit. Applying transparent lean standards to the audit itself, the audit report should be transparent, comprehensive, and accurate. The lean audit report should be presented to management and all involved stakeholders. It must contain an executive summary, the result, and a detailed action plan, including corrective action requests (CARs). CARs should be linked with a clear responsibility (name, role, and function) and a closing date. The audit report should be signed by the lean auditors and the top management to obtain a commitment for the continual closing of identified improvement areas (Helmold *et al.*, 2023).

Phase four is the follow-up after the audit. The best way to ensure that companies are effectively utilizing lean methodology is to follow up with management, each area manager, and stakeholders after the audit. This includes making sure that any relevant issues have been addressed and noting what standards still need to be met. A weekly self-checking audit or monthly audit carried out by a supervisor can help your company meet and maintain lean standards in the future. The follow-up is normally conducted after three months or six months, accompanied by self-checks and progress reports.

12.1.3 5S Audits

The 5S framework, developed and popularized in Japan, provides five key steps for maintaining an efficient workspace in order to improve the quality of products. In Japanese, these steps are known as Seiri (sort), Seiton (set in order), Seiso (shine), Seiketsu (standardize), and Shitsuke (sustain). Work Aid 02 (5S Audit Checklist) can be used as a practical tool for assessing waste within a certain area. Figure 12.3 shows the example of a 5S Audit at the BKSB in Berlin.

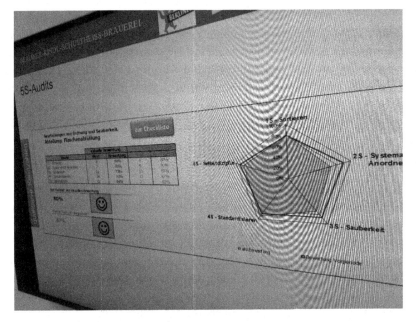

Figure 12.3.   5S Audit at the BKSB.

12.1.4 Go Gemba Audits (GoGA)

A Go Gemba Audit or Gemba Walk is a systematic and regular workplace-related assessment that aims to identify and eliminate waste by asking employees, observing their tasks, evaluating the processes, and scrutinizing work routines. Gemba Walk is derived from the Japanese word "Gemba," or which means "the real place," so it is often literally defined as the act of seeing where the actual work happens (Helmold, 2023).

12.1.5 Value Stream Audits (VSA)

Value stream audit, also known as value stream mapping (VSM), is a lean-management method for analyzing the current state and designing a future state for the series of events that take a product or service from the beginning of the specific process until it reaches the customer. The original VSM template was created by Toyota Motor Company and implemented via material and process flowcharts. This VSM illustrated the necessary process steps that existed from order entry to final product delivery and was useful for gaining a wide-reaching view of the company's activities (Helmold, 2023). It allowed Toyota to remove non-essential activities that created waste while maintaining the manufacturing process.

12.1.6 Lean Audit Workshop (LAW)

A lean manufacturing workshop gives companies a comprehensive introduction to lean concepts and tools. Workshops can be based on their scope, lasting between three days and six months. Lean audit workshops focus on implementing a pull, zero defect, flow, and tact principle within the value chain. Lean audit workshops target reducing any waste of time, effort, or money by recognizing each step in a business process and then restructuring or cutting out steps that do not create value. It creates more value for customers with fewer resources. Lean in simple terms means reducing unwanted activities, processes, or anything else that does not add value to the product or service for the customer. Guiding principles for lean management include:

- Defining value for the customer.
- Identifying each step in a business process.
- Eliminating those steps that do not create value.
- Making the value-creating steps occur in an efficient sequence.
- Repeating the above steps on a continuous basis until all waste has been eliminated.

12.2 Audits

12.2.1 Scope and Definition

Audits can be described as a systematic and structured performance evaluation and assessment of a system, process, product, or any other area by internal or external auditors. The aim of an audit is to evaluate and approve or disapprove the assessed area using standardized criteria and questions, to define areas for action, and to ensure the sustainable implementation of the actions and improvement areas. Assessment criteria in audits are based on customer and stakeholder expectations. Audits can be clustered into systems, processes, products, controls, and special audits, as shown in Table 12.1. Lean audits are conducted to determine if the business is properly implementing lean management methodologies into the company and value chain (Helmold & Terry, 2021). This is achieved by a detailed 360° analysis of lean processes with the goal of recognizing opportunities to improve processes and to eliminate waste (Helmold *et al.*, 2023).

Table 12.1.   Audit types.

Audit type	Description
Systems Audit	Evaluation of the (Quality Management) System of the Organization by external certification agencies (TÜV, DEKRA, Bureau Veritas). Examples: DIN EN ISO 9001:2015, IATF 16949, International Railway Industry Standard (IRIS).
Process Audit	Evaluation of a (Manufacturing or Service) Process (Input-Transformation-Output) to qualify or disqualify a process-oriented example of a product or service by assessing a reference process from supply side, incoming material to the dispatch (also from other customers). Examples: VDA 6.3, SEAP (Supplier Evaluation Approval Process — Railway).
Product Audit	Planning and execution of the assessment of a finished product to be delivered to the customer. The audit consists of checking the specification, drawings, capacity, and other important aspects and normally involves the trial run of the entire manufacturing process (e.g., 300 parts, run at rate). Examples: VDA 6.5, Part Production Approval Process (PPAP), Production Approval Process (PAP).
Control Audit	Control audits (normally after 1, 3, or 6 months) aim to control (verify or falsify) the progress of previously conducted audits. Control Audits normally target reducing the previously found Corrective Action Requests (CARs).
Other Audits	Any other audits in areas like safety, health, environment, tax, and financials. Examples: 5S audits, tax audits, environmental audits (ISO 14001), IT audits (ISO 27001), financial audits, or health, safety, and environment (HSE) audits.

12.2.2 Systems Audits

The system audit evaluates the standard requirements for quality management systems (QMS). The auditing of a management system, e.g., DIN EN ISO 9001:2015, is referred to as a system audit. This can also be a combination of several management systems, such as environment, quality, and occupational safety, which are then referred to as an integrated management system. DIN EN ISO 9001 specifies the minimum requirements for a QMS for the manufacture of products or services (QM system; QMS), which an organization must meet in order to be able to provide products and services that meet customer expectations and any official requirements. At the same time, the management system should be subject to a continuous improvement process. Although the process-oriented approach was already introduced with the 2000 revision, there were considerable problems with its implementation. This should be made easier by the revision. The standard also calls for a more risk-based approach. A formal QM manual will no longer be necessary if the organization otherwise provides adequate documentation. There will also no longer have to be a "top management representative" in the formal sense. The current version of ISO 9001 was last revised in 2015. Based on EN ISO 9001, IATF 16949 exists for series production in the automotive industry. Compared to EN ISO 9001, it places more extensive requirements on the QMS. The basic idea of ISO 9001:2015 is that for long-term success, companies must consider the requirements of their stakeholders (Brugger-Gebhardt, 2016). That is why the standard emphasized the interested parties even more as a separate point. In contrast to ISO 9001:2008, the focus is no longer just on the customer but on the interest groups (interested parties). In addition to customers, this includes suppliers, owners, employees, authorities, business partners, or even competitors. ISO 9001 continues to follow the approach of planning (plan), performing (do), checking (check), and acting (act), or PDCA cycle for short, in order to continuously and effectively improve the QMS as a whole and its processes (Weidner, 2016).

The ISO 9001 will be reviewed by the audit committee to get a revision. Accordingly, "adjustments to ISO 9001 are planned with a view to the aspects of resilience, supply chain management, change management, sustainability, dealing with risks [and] knowledge of the organization."

However, the basic structure of the standard should be retained. The revision is simply intended to replace the previously used High Level

Structure (HLS) with the new Harmonized Structure (HS), as was already done when revising the new ISO/IEC 27001:2022.

The decision to revise the ISO 9001 standard "early" is presumably due to current changes in the company environment, such as increasing complexity and dynamism and the use of new technologies, which make adjustments with regard to the application of QMS urgently necessary. The next steps include the usual procedure for revisions of ISO management system standards, such as the establishment of working groups and the appointment of project managers (DGQ, 2023).

12.2.3 Process Audits

A process audit is used in supplier management to assess the quality and performance of input factors, processes, and their outputs (Helmold *et al.*, 2023). The process audit is a central part of supplier management and is intended to lead to capable and controlled processes at suppliers that are robust to disruptions. By stabilizing the processes, product quality is increased and complaints are prevented (Helmold, 2021). The following objectives are pursued with process audits:

Prevention: Recognizing, pointing out, and initiating measures that prevent deficits from occurring.
Correction: Analyzing known deficiencies and taking actions to correct and prevent recurrence.
Continuous improvement process: Further improving implemented measures from a process audit to make the process more capable and robust.

Process audits can be applied throughout the product life cycle. This includes internal and external processes, e.g., the entire supply chain. By using process audits, possible process risks in the entire product development process (PEP) can be identified at an early stage. Process audits evaluate company process chains from input through transformation to output. Due to the process-oriented approach, the standard is easy to apply and independent of the size of the company and the purpose of the company. The effectiveness and efficiency of the organization in achieving the set goals are improved, and customer satisfaction is also increased by meeting expectations. A process is a set of interrelated or interacting activities that converts inputs into results (output). The process approach enables an organization to understand requirements better and meet them more consistently (improved, consistent, and predictable outcomes). A product audit focuses:

- On quantitative and qualitative characteristics;
- On physical products and related processes;
- On each production step until final completion;
- On the transfer to the next customer (internal/external);
- On achieving the targeted state and specification;
- On meeting expected criteria.

12.2.4 Product Audits

As part of the product validation and testing of a defined number of products, the product audit confirms the quality capability of the production process based on the quality characteristics of a product. It is checked whether the product corresponds to the specified specifications, special customer, and supplier agreements. A product audit is the planning, implementation, evaluation, and documentation of tests (Helmold, 2023).

A product audit is used to assess compliance with the company's own quality requirements. In addition, it aims to assess compliance with the expressed and unspoken customer requirements (with the "eyes of a very critical customer"). The product audit represents a measure to check the effectiveness of quality. Checks and control measures are carried out that lead directly to process and product improvements in a short time. Finally, it creates an internal basis of trust with regard to the requirements of product liability and checks the conformity of the products, including any legal requirements. Within the automotive industry, the Production Part Approval Process (PPAP) is a common product qualification process (Helmold, 2022). It is a procedure from the QS 9000, which has now been replaced by ISO/TS 16949, in which series parts are sampled. This procedure originates from the automotive industry and has been successfully implemented for years. It's all about the quality of the parts supplied, which means that the parts from the series tools or series processes must correspond to the drawings. In addition to the parts delivered for inspection, the part submission warrant (PSW) is a central element of the sampling process.

12.2.5 Control Audits

The control audit is a special type of audit outside the regular audit plan within the value chain to verify and control the progress of audits and can have the following reasons:

- Progress control;
- Special process audits, e.g., for processes such as gluing, painting, and welding;
- Escalating an audit;
- Audits based on customer requests.

12.2.6 Other Audits

Other audits include all possible assessments of standard requirements in sub-areas through 5S audits, environmental audits, financial audits, security audits, etc. Other audits can be the environmental standard ISO 14001, IT standards like ISO 27001, or others (Heras-Saizarbitoria, 2018).

12.3 International Organization for Standardization

The International Organization for Standardization (ISO) is an independent, non-governmental international organization with a membership of 167 national standards bodies (Geiger & Kotte, 2008). Through its members, it brings together experts to share knowledge and develop voluntary, consensus-based, and market-relevant international standards that support innovation and provide solutions to global challenges.

Full members (or member bodies) influence ISO standards development and strategy by participating and voting in ISO technical and policy meetings. Full members sell and adopt ISO International Standards nationally. Correspondent members observe the development of ISO standards and strategy by attending ISO technical and policy meetings as observers. Correspondent members that are national entities sell and adopt ISO International Standards nationally. Correspondent members in territories that are not national entities sell ISO International Standards within their territory. Subscriber members keep themselves up to date on ISO's work but cannot participate in it. They do not sell or adopt ISO International Standards nationally.

12.4　Definition and Scope of QMS

A QMS is the combination of methods, principles, and processes of quality excellence applied in an organization. A QMS always focuses on meeting and exceeding customer requirements. It has a set of guidelines that are defined by a collection of policies, processes, documented procedures, and records. This system defines how a company will achieve the creation and delivery of the product or service they provide to their customers (Pfeiffer & Schmidt, 2014). When implemented in your company, the QMS needs to be specific to the product or service you provide, so it is important to tailor it to your needs. However, in order to help ensure that you do not miss elements of a good system, some general guidelines exist in the form of ISO 9001 (QMS — Requirements), which is intended to help standardize how a QMS is designed. ISO 9001 is the international standard for QMS, published by ISO. The International Organization for Standardization (ISO) is an independent, non-governmental, international standard development organization composed of representatives from the national standards organizations of member countries. The major aim is to apply internationally harmonised quality standards. The standard was most recently updated in 2015 and is referred to as DIN EN ISO 9001:2015. In order to be released and updated, ISO 9001 had to be agreed upon by a majority of member countries so that it would become an internationally recognized standard, which means it is accepted by a majority of countries worldwide. ISO has a range of standards for QMS that are based on ISO 9001 and adapted to specific sectors and industries. QMS is accredited by globally applied standards. The advantages of a QMS can be outlined as follows:

- Increasing customer satisfaction by using a globally applied standard and improvement system.
- Becoming more cost-efficient, increasing credibility, and securing competitiveness.
- Optimizing costs and creating shorter cycle times through effective use of resources.
- Enhanced customer satisfaction and improved customer loyalty leading to repeat business.
- Increased revenue and market share obtained through flexible and fast responses to market opportunities.

- Integration and alignment of internal processes, which will lead to increased productivity and results.
- Ensuring a consistent and streamlined delivery of the products or services requested by customers.
- Improved communication, planning, and administration processes throughout the organization.

12.5 Different Types of QMS

12.5.1 Scope and Definition

A QMS describes in enterprises and organizations the management function and all organizational activities which serve the improvement of the process quality, the work quality, and thus product and service quality. QMS uses lean features for process improvements. Table 12.1 outlines the most common standards of QMS in certain industries.

Audit types and the relevant sectors are shown in Table 12.2.

12.5.2 QMS: DIN EN ISO 9001:2015

The DIN EN ISO 9001:2015 (ISO 9001) QMS is the world's most popular quality management standard, with over one million certified organizations in 180 countries worldwide. ISO 9001 provides a quality management framework that companies can use to ensure the quality of their products and services is consistent. Companies choose ISO 9001 certification to show that they have taken due care to maintain high standards. This reduces the chance of product faults, recalls, or service shortcomings and

Table 12.2. Quality management systems (QMS) and sectors.

Name	Industry
DIN ISO 9001:2015	General Quality Management System
EN/AS 9100	Aviation Sector
IATF 16949:2016	Automotive Sector
DIN EN 15224	Healthcare Sector
ISO/TS 22163	Railway, replacing IRIS
ISO 13485	Medical Industry
TL 9000	Telecommunications

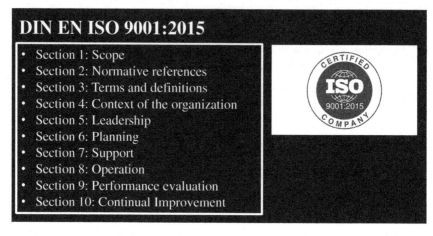

Figure 12.4.   DIN EN ISO 9001:2015.

ensures that customers can buy with confidence. ISO 9001 certification demonstrates an organization's ability to consistently meet and exceed customer expectations. Many enterprise buyers and retailers require their suppliers to be ISO 9001 certified in order to minimize their risk of purchasing a poor product or service — being ISO 9001 certified boosts the sales potential. A business that achieves ISO 9001 certification can attain significant improvements in organizational efficiency and product quality by minimizing waste and errors and increasing productivity. Our comprehensive guideline on ISO 9001 provides useful information on requirements and benefits. Figure 12.4 depicts the DIN ISO 9001:2015 elements. The elements of ISO 9001:2015 are:

Section 1: Scope
Section 2: Normative references
Section 3: Terms and definitions
Section 4: Context of the organization
Section 5: Leadership
Section 6: Planning
Section 7: Support
Section 8: Operation
Section 9: Performance evaluation
Section 10: Continual Improvement

12.5.3 Aviation QMS: EN/AS 9100

Aviation thrives on safety like hardly any other industry. Therefore, QMS according to EN/AS 9100ff is becoming increasingly important in the aviation industry. It enables the company to meet all quality requirements and to continuously improve. ISO 9001 forms the basis for the formulation of EN/AS 9100ff. Certification according to EN/AS 9100ff therefore also includes certification according to ISO 9001. The EN/AS 9100 series of standards includes the following requirements in the appendix:

- A completely traceable recording procedure over the entire supply chain of a product.
- Definition of the interfaces to customers and aviation authorities in the procedures.
- Definition, implementation, and documentation of verification and validation activities.
- Execution, implementation, and documentation of initial sample tests.

12.5.4 Automotive QMS: IATF 16949:2016

The International Automotive Task Force (IATF) is a group of automotive manufacturers which aims at providing improved quality products to automotive customers worldwide. The standard IATF 16949 contains general requirements for QMS of mostly the North American and European automotive industries. They were developed jointly by the IATF members and published based on EN ISO 9001. About 30% of the more than 100 existing car manufacturers agree with these harmonized requirements of the seven IATF members (BMW, Daimler, Ford, General Motors, Renault, Stellantis, and VW) — but the large Asian car manufacturers in particular have differentiated and have their own requirements for the QMS of their group and suppliers. IATF16949:2016 can be applied throughout the automotive supply chain. Certification is based on the certification specifications (rules) issued by the IATF. The certificate is valid for three years and must be confirmed annually by IATF-certified auditors (third-party auditors) from accredited certification companies. This is followed by re-certification for a further three years with renewed annual confirmation. IATF16949:2016 must not be regarded as an independent QMS standard but is to be understood as a supplement to ISO 9001:2015. Certifications must therefore include both standards. The certification may only be carried out by bodies that have been approved by the IATF (so-called certification bodies). A certificate, according to IATF16949:2016, should justify the trust of the (potential) customer in the system and process quality of a (possible) supplier. Today, a supplier without a valid certificate has little chance of supplying a Tier 1 automotive supplier and certainly not an automotive manufacturer (OEM) with series parts. Certification according to the previous ISO/TS 16949 standard was only possible until September 2017. Since 1 October 2017, certifiers are only allowed to carry out audits and issue certificates according to the new standard IATF 16949. This applies to initial audits, surveillance audits, and recertification audits. The previous certificates according to ISO/TS 16949 lost their validity in September 2018.

The 10 chapters of the IATF 16949 standard (HLS) are:

Chapters 0–3: Introduction, Scope, Normative references, Terms
Chapter 4: Context of the organization
Chapter 5: Leadership
Chapter 6: Planning
Chapter 7: Support
Chapter 8: Operation
Chapter 9: Evaluation of performance
Chapter 10: Improvement

12.5.5 Railway QMS: ISO/TS 22163 (IRIS)

At the end of May 2017, the Association of the European Rail Supply Industry Association (UNIFE) announced the release of the new global standard for QMS in the rail industry, ISO/TS 22163, by the Technical Committee 269 of the ISO. ISO/TS 22163 is based on the structure and requirements of the ISO 9001:2015 standard and also includes additional requirements for the rail industry. Within the timeframe of the defined transition phase, ending in September 2018, organizations certified according to IRIS Rev.02 will have to undergo transition audits to update their certificates to ISO/TS 22163. As IRIS Rev.02 certificates will expire after 14 September 2018, and all non-conformities need to be closed for all audits before a new ISO/TS 22163 certificate can be issued, the last transition audits should be completed by July 2018. The International Railway

Industry Standard (IRIS) is based on the ISO 9001 standard and is tailored to the requirements of the international rail industry. IRIS certification ensures uniformity of language and guidelines for evaluation, as well as mutual acceptance of audits, ensuring a high level of transparency across the supply chain. IRIS is applicable to companies engaged in the manufacture and/or design and/or maintenance of rolling stock and signalling systems: equipment manufacturers, system integrators, operators, or business partners. IRIS has been developed by the UNIFE with the aim of becoming a commonly recognized system for assessing the quality of rolling stock suppliers and related equipment and components, internationally recognized.

For sales in Europe, IRIS certification is mandatory. The certification procedure is as follows:

- Organization registers for membership at the UNIFE portal.
- Compilation of the information questionnaire.
- Pre-audits to verify potential gaps in compliance with the IRIS standard.
- Readiness Review audits to verify compliance with IRIS prerequisites.
- Certification audit.
- Issue of the certification.
- Annual supervision audits.
- Recertification audit (after three years).

The successful implementation of the IRIS Certification creates a win–win situation for all stakeholders in the rail sector, be they equipment manufacturers, system integrators, operators, or business partners. More broadly, the IRIS Certification scheme helps to develop the attractiveness of rail, shifting our daily mobility toward more sustainable modes of transportation. ISO/TS 22163:2017 is the standard specifying business management system requirements for rail organizations. ISO/TS 22163:2017 is largely based on ISO 9001:2015, which is a general QMS standard used by millions of companies around the world. In addition, ISO/TS 22163:2017 includes specific requirements tailor-made for high-quality business management in the railway sector. The standard functions as a formula describing the best practices for business management in the rail sector.

12.5.6 Healthcare QMS: DIN EN 15224

Healthcare providers today rely on ISO 9001 QMS as a framework for success. However, in order to truly achieve organizational efficiency, drive productivity, and increase profitability, the continued reliance on ISO 9001 is insufficient. Healthcare providers seeking to differentiate themselves have to ensure that the core health-related issues such as patient safety and management of clinical risks are addressed and properly handled as well. This particular point is not covered under the ISO 9001 standards. The DIN EN 15224 QMS is a sector-specific standard of quality management for healthcare organizations. It is based on the ISO 9001 standard, and it includes tangible requirements for patient safety and management of clinical risks in planning, realization, and management processes.

The DIN EN 15224 system brings together the advantages of ISO 9001 with comprehensive healthcare quality requirements. It defines issues ranging from the effectiveness and suitability of care to the safety and reliability of care processes. In DIN EN 15524, there are 11 quality features which characterize the quality of healthcare and must be verified as part of a certification. The 11 central features of DIN EN 15224 are:

(1) Appropriate, correct care
(2) Availability
(3) Continuity of the care
(4) Effectiveness
(5) Efficiency
(6) Consistency
(7) Evidence-based, knowledge-based care
(8) Patient care, including physical and psychological integrity of the care
(9) Integration of the patient
(10) Patient safety
(11) Timeliness and accessibility

Compliance with the DIN EN 15224 QMS allows your business to commit to the long-term goal of providing the best healthcare services. By adhering to the requirements for the certification, not only do you increase organizational efficiency and product quality but also minimize waste and mistakes, thereby increasing the overall productivity.

12.5.7 Medical QMS: ISO 13485

QMS elements of ISO 13485 in the medical industry is crucial for safety and security reasons. ISO 13485 is a harmonized standard which lays down the requirements for QMS for medical devices. Medical device manufacturers have to therefore, above all, be certified according to ISO 13485, because according to Appendix II of the Medical Device Directive (MDD), they can explain the compliance of their products themselves. For medical devices that incorporate software or standalone software, IEC 62304 also demands a QMS and recommends ISO 13485. The validity of the QMS will be examined by external auditors (usually notified bodies) and internal audits.

12.5.8 Telecommunications QMS: TL 9000

The telecommunications industry is a fast-growing industry and has established TL 9000 as a QMS in this industry. TL 9000, based on the ISO 9001 system, is a QMS designed to meet supply chain requirements of the telecommunication industry. The system ensures quality by defining system requirements central to the design, development, production, delivery, deployment, and maintenance of telecommunications products and services. It also provides a measurement system for tracking performance and improving results. TL 9000 certification eliminates the need to multiple quality management standards, making it a one-stop solution for telecommunications industry providers who are looking to demonstrate their commitment to quality standards. Adopting the TL 9000 management system allows your company to demonstrate its compliance with the high standards required by the information communication technology (ICT) sector and telecommunications industry. It also aids in the validation of the quality of the processes, products, and services. TL 9000 enables the company to create value for customers and stakeholders through continuous improvement, strategic partnerships, and high standards achieved through a solid measurement system.

References

Brugger-Gebhardt, S. (2016). *Die DIN EN ISO 9001:2015 verstehen: Die Norm sicher interpretieren und sinnvoll umsetzen*. Wiesbaden: Springer.

DGQ. (2023). DGQ. Revision ISO 9001:2015. https://www.dqsglobal.com/de-de/wissen/blog/revision-iso-9001 (Accessed April 15, 2024).

Helmold, M. (2021). *Successful Management Strategies and Tools. Industry Insights, Case Studies and Best Practices.* Cham: Springer.

Helmold, M. (2022). *Strategic Performance Management. Achieving Long-term Competitive Advantage through Performance Excellence.* Cham: Springer.

Helmold, M. (2023). *Erfolgsformel LeanMit Kaizen, Kata und Keiretsu Wettbewerbsvorteile erzielen.* München: Hanser Verlag.

Helmold, M. & Terry, B. (2021). *Operations and Supply Management 4.0 Industry Insights, Case Studies and Best Practices.* Cham: Springer.

Helmold, M. et al. (2023). *Qualität neu denken. Innovative, virtuelle und agile Ansätze entlang der Wertschöpfungskette.* Wiesbaden: Springer.

Heras-Saizarbitoria, I. (2018). *ISO 9001, ISO 14001, and New Management Standards.* Cham: Springer.

IATF 16949:2016 Anforderungen an Qualitätsmanagementsysteme für die Serien- und Ersatzteilproduktion in der Automobilindustrie 1. Ausgabe, Oktober 2016. VDA, abgerufen am 16. März 2018.

Paterson, J. C. (2015). *Lean Auditing: Driving Added Value and Efficiency in Internal Audit.* London: Wiley.

Pfeiffer, T. & Schmidt, R. (2014). Qualitätsmanagementsysteme, Teil II in: Masing Handbuch Qualitätsmanagement. Carl Hanser Fachbuchverlag München Wien, 6. überarbeitete Auflage.

Weidner, G. E. (2016). *Qualitätsmanagement: Kompaktes Wissen — Konkrete Umsetzung — Praktische Arbeitshilfen.* München: Hanserverlag.

Chapter 13

Realities (Mixed, Augmented and Virtual Realities) in Lean Management

13.1 Extended Realities (XR)

Extended reality (XR) is an umbrella term encapsulating augmented reality (AR), virtual reality (VR), mixed reality (MR), and everything in between. Although AR and VR offer a wide range of revolutionary experiences, the same underlying technologies are powering XR. XR applications are used in industries such as healthcare, defense, education, construction, engineering, gaming, and manufacturing (Lang & Müller, 2020). Realities enable multidimensional content viewing and collaboration throughout the value chain through cloud-based systems and in real-scale environments, resulting in better processes and higher levels of efficiency. Realities help to eliminate waste in all stages of a product life-cycle and offer therefore an ideal and integrative lean management approach that combines project management philosophies, systems, technologies, and tools (Alizadehsalehi & Hadavi, 2023). XR advances the potential of AR, VR, and MR and merges our real and virtual worlds to create new environments and visualizations where physical and digital objects co-exist, interact, and communicate (Marr, 2021). Instead of removing users completely from the real world or simply layering flat content on top of their immediate view, MR adds intelligence, even personality, to digital content relative to the world around them (Helmold, 2024). As part of the technological evolution of how we engage with the digital world, in both our personal and work lives, we are smashing through the barriers that interfere with our ability to make smart decisions quickly, absorb, retain, and process critical information, visualize possible scenarios before acting or share knowledge and tasks. The XR trend is ushering in a new world of simulated experiences grounded in the ways business gets done and how customers actually use products (Fritz, 2022). AR or XR enables a gradual full or partial virtual representation of reality, which can either closely resemble reality or completely deviate from it. The boundaries between reality and the virtual world are becoming increasingly blurred thanks to new hardware and computing power as well as high-performance networks with high bandwidth, opening up new, breathtaking impressions that until recently could only be imagined in science fiction films. AR is also a key technology for what is known as the Metaverse, the third major generation of the Internet, also known as Web 3.0. This follows on from documents and their linking based on the first generation and the second generation shaped by social media. AR systems have been researched and developed for quite some time. The first systems for computer games were already commercially available in the 1990s. However, only in the last 10 years have systems been developed that enable good performance while remaining cost-effective. The simplest systems consist of cardboard constructions into which the smartphone can be installed. Figure 13.1 shows a first overview of the XR

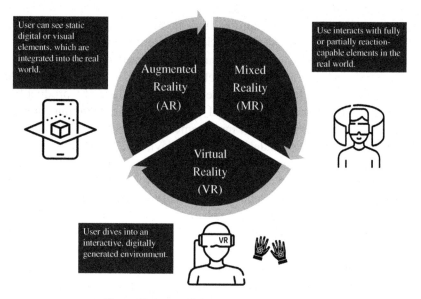

Figure 13.1.   Realities in lean management.

systems (VR, MR, and AR), which are described in detail in the following sections.

13.2 Creation of XR

The realities mentioned above are generated by suitable hardware. First of all, these include glasses, which are also referred to as head-mounted displays (HMDs). There are three categories: connected (tethered), wireless (mobile), and fully autonomous (standalone) devices:

- *Tethered*: These glasses must be tethered and connected to a powerful computer system.
- *Wireless (mobile)*: This solution consists of an attachment with which smartphones approved for this purpose can be used. A cable connection is no longer required, which significantly increases the range of use and enables new applications.
- *Autonomous (standalone)*: Devices that do not require a connection to a computer or the use of a smartphone are referred to as autonomous or standalone. They offer the greatest possible flexibility and the implementation of demanding applications.

Regardless of this category, the following parameters are important when choosing a suitable solution:

- *Display*: resolution and type (AMOLED, OLED, LCD, etc.)
- Resolution (per eye)
- Connections (USB, Bluetooth)
- Position tracking
- Field of view

In standalone systems, the processor (e.g., Intel, Snapdragon, or Nvidia) and the platform (Windows MR, Google Daydream) play a particularly important role.

Of course, different systems are possible for VR applications and AR applications. HTC Vive and Oculus Rift are two leading providers of tethered VR-headsets on the market (Helmold, 2024). Microsoft HoloLens is a leading system for AR applications. For the realization of AR on mobile phones, Apple's ARKit and Google's ARCore are very widespread and powerful systems.

In order to bring the virtual world closer to reality, HMDs can be supplemented with wearable elements such as tactile gloves, suits, or vests. The impression in the virtual world is thereby again clearly strengthened. Haptic gloves enable the position of fingers and hands to be recorded, thereby creating another interface for control between humans and machines. Such gloves also enable feedback such as vibrations, impressions of force, or even the feeling of surface structures. Holding and interacting with objects can be simulated with controllers.

Only the combination of these hardware elements with suitable software then enables the creation of a suitable XR system.

The current forecasts for market growth from $12 billion in 2020 to over $72 billion in 2024 show the great potential of this technology (cf. Helmold, 2024). Of course, this development is supported by the gaming market, but these systems are also increasingly found in production environments. However, some prerequisites must be created for their smooth use. This includes, for example, sufficient lighting conditions or the detection and removal of highly reflective parts.

13.3 Augmented Reality

AR is an interactive experience that combines the real world and computer-generated content. The content can span multiple sensory modalities, including visual, auditory, haptic, somatosensory, and olfactory. AR incorporates three features: a combination of digital and physical worlds, interactions made in real time, and accurate 3D identification of virtual and real objects. Quality assurance could be done in real time by a human or even by an AI with AR applications, as shown in Figure 13.2. Either AR glasses could place 3D models on the products, or the comparison can be left to an AI.

13.4 Mixed Reality

MR enables interaction with and manipulation of digital or virtual and real objects in real time. MR is sometimes also referred to as hybrid reality and makes it possible to interact with virtual objects. MR requires a suitable headset (e.g., Microsoft HoloLens) and significantly more computing power than VR or AR. Virtual objects can then be placed in space, rotated, and further interacted with. MR applications can also be implemented on

Figure 13.2.   Augmented reality.
Source: Helmold, M. (2024). Wiesbaden: Springer.

smartphones or tablets, although the boundaries between AR and MR are becoming increasingly blurred here. One example is Google 3D Experience, which allows virtual objects such as animals to be placed and scaled in real space. This can be used for training, for example. It is also possible to get an idea of how certain objects, such as furniture, fit into a room.

13.5 Virtual Reality

VR is a simulated experience that employs pose tracking and 3D near-eye displays to give the user an immersive feel of a virtual world. VR is the use of computer technology to create a simulated environment which can be explored in 360°. Unlike traditional interfaces, VR places the user inside the virtual environment to give an immersive experience. VR enables users to explore and interact with a virtual surrounding in a way that approximates reality, as it is perceived through the users' senses. The environment is created with computer hardware and software, although users might also need to wear devices such as helmets or goggles to deeply immerse themselves in a VR environment.

The VR industry still has far to go before realizing its vision of a totally immersive environment that enables users to engage multiple sensations in a way that approximates reality. However, technology has come a long way in providing realistic sensory engagement and shows promise for business use in a number of industries.

VR systems can vary significantly from one to the next, depending on their purpose and the technology used, although they generally fall into one of the following three categories:

- *Non-immersive*: This type of VR typically refers to a 3D simulated environment that's accessed through a computer screen. The environment might also generate sound, depending on the program. The user has some control over the virtual environment using a keyboard, mouse, or other device, but the environment does not directly interact with the user. A video game is a good example of non-immersive VR, as is a website that enables a user to design a room's decor.
- *Semi-immersive*: This type of VR offers a partial virtual experience that's accessed through a computer screen or some type of glasses or headset. It focuses primarily on the visual 3D aspect of VR and does not incorporate physical movement in the way that full immersion does. A common example of semi-immersive VR is the flight simulator, which is used by airlines and militaries to train their pilots.
- *Fully immersive*: This type of VR delivers the greatest level of VR, completely immersing the user in the simulated 3D world. It incorporates sight, sound, and, in some cases, touch. There have even been some experiments with the addition of smell. Users wear special equipment such as helmets, goggles, or gloves and are able to fully interact with the environment. The environment might also incorporate such equipment as treadmills or stationary bicycles to provide users with the experience of moving through the 3D space. Fully immersive VR technology is a field still in its infancy, but it has made important inroads into the gaming industry and, to some extent, the healthcare industry, and it's generating a great deal of interest in others.

References

Alizadehsalehi, S. & Hadavi, A. (2023). Synergies of lean, BIM, and extended reality (LBX) for project delivery management. *Sustainability*, 15(6), 4969.

Fritz, J. (2022). *Datenbasierte Optimierung des Business Management Systems — Geschäftsprozesse verbessern mit Data Analytics, Industrie 4.0, KI*. München: Chatbots und Co. Hanser Verlag.

Helmold, M. (2024). *Erfolgreiche Transformation zum digitalen Champion: Wettbewerbsvorteile durch Digitalisierung und Künstliche Intelligenz*. Wiesbaden: Springer.

Lang, M. & Müller, M. (2020). *Von Augmented Reality bis KI — Die wichtigsten IT-Themen, die Sie für Ihr Unternehmen kennen müssen*. München: Hanserverlag.

Marr, B. (2021). *Extended Reality in Practice: 100+ Amazing Ways Virtual, Augmented and Mixed Reality Are Changing Business and Society*. Chennai: Wiley.

Chapter 14

Lean Throughout the Value Chain — Keiretsu

14.1 Scope and Definition

A Keiretsu network (系列 ネットワーク), or *keiretsu* value chain network (Japanese: integration, order or system of stakeholders, partners, and suppliers), represents a means of mutual security, especially in Japan, and usually includes large manufacturers and their suppliers of raw materials, systems, and components (Ahmadin & Lincoln, 2001). Keiretsu groups are defined as clusters of independently managed firms maintaining close and stable economic ties, cemented by a governance mechanism such as presidents' clubs, partial cross-ownership, and interlocking directorates. Within the broad definition lie two distinctive variations. The horizontal *keiretsu* are conglomerates covering several industries linked by cross-shareholding, intra-group financing, and high-level management by a central (often inconspicuous) body of directors. The vertical *keiretsu* are grouped around one big manufacturer and consist of a multilayered system of suppliers focused on the core company (Gabrowiecki, 2006). Keiretsu networks have received much attention in the European automotive and transportation sector through the success of Japanese companies like Toyota, Mitsubishi, or Hitachi and other conglomerates in achieving improved customer service, better inventory control, and more efficient overall channel management (Freitag, 2004). Keiretsu, which is a form of Japanese business network, shares many of the goals of several business functions. The concept of a *keiretsu* network was introduced by Toyota in the mid-1980s (Imai, 1986) and transferred to affiliates and suppliers

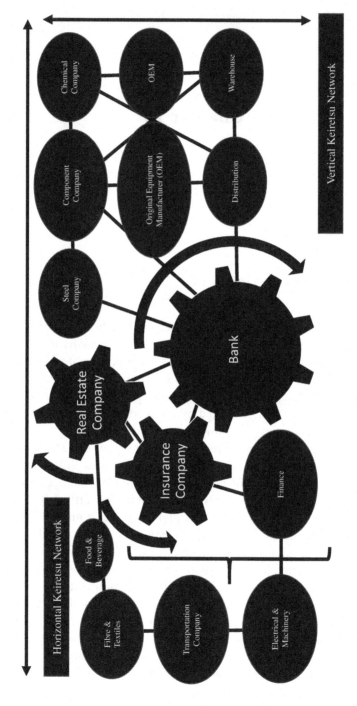

Figure 14.1.   Keiretsu network structure.

outside Japan (Kalkowsky, 2004). Figure 14.1 depicts the example of a Keiretsu network structure, including the core business functions and enterprises of bank, insurance, and finance companies (Gabrowiecki, 2006). These core businesses are surrounded by companies in leading industries such as automotive, food or machinery, and electrical applications. In addition to this, *keiretsu* networks may include supporting businesses such as warehousing, transportation, or component supplies. Keiretsu networks are differentiated into horizontal (same layer) and vertical *keiretsu* networks (different layers) (Hoshi, 1994).

Keiretsu networks often include partial ownership of the respective supplier as collaborative *keiretsu* or supply networks (Shun Cai *et al.*, 2013). Control relationships between pairs of firms represent a form of bilateral exchange. The school of *keiretsu* may lead to broad functional and cultural changes for those companies which use the system (Freitag, 2004). Keiretsu networks with financial and commercial connections develop quasi-administrative ties through cross-shareholding, as Keiretsu networks have two sides: (1) horizontal relationships based on mutual support, and (2) vertical structures based on asymmetric exchange and control between financial firms and industrial firms. In various articles and books, Liker explains the Toyota way and the principles of *keiretsu* supply networks (Liker, 2004). Many original equipment manufacturer (OEMs) and their suppliers have meanwhile adopted this system (Liker & Choi, 2005).

- Keiretsu refers to the Japanese business structure comprising a network of different companies, including banks, manufacturers, distributors, and supply chain partners (Shimotani, 1995).
- Before the *keiretsu* system, the primary form of corporate governance in Japan was the *zaibatsu*, which referred to small, family-owned businesses that eventually evolved into large, monopolistic holding companies.
- A horizontal *keiretsu* refers to an alliance of cross-shareholding companies led by a Japanese bank that provides a range of financial services.
- A vertical *keiretsu* is a partnership of manufacturers, suppliers, and distributors that work cooperatively to increase efficiency and reduce costs.
- A drawback of the *keiretsu* system is the easy access to capital, which can lead a company to take on too much debt and invest in risky strategies.

14.2 Origin of Keiretsu Networks

It is widely accepted that the formation of enterprise groups called *keiretsu* was established in Japan after the Second World War by Toyota and other huge enterprises. On the other hand, some authors indicate that the roots of the above-mentioned groups were in the Meiji period. After Japan had come out of its two-century policy of isolation, the young Meiji government created the initial infrastructure for future industrialization. Moreover, it was actively involved in establishing and expanding companies. The result of this policy was the creation of a *zaibatsu* (Kensy, 2001). The *zaibatsu* were family-run enterprise groups. Morikawa (1992) defines *zaibatsu* "as a group of diversified business owned exclusively by a single family or an extended family." Zaibatsu companies started to develop themselves very quickly because they obtained a lot of subsidies and contracts from the Meiji government. The first *zaibatsu* to be created was Mitsui, which was established in 1876. The next three established *zaibatsu* were: Mitsubishi, Sumitomo, and Yasuda. Mitsubishi was concentrated on shipbuilding and heavy industry and was a major player in mining, shipping, trade, brewing, insurance, and banking. Sumitomo was not only focused on banking but also on mining and metals. Yasuda *zaibatsu* specialized in finance, controlling an important bank, a major trust bank, and two large insurance firms. These four *zaibatsu* were called the Big Four. They expanded their financial business, also establishing insurance companies and trust banks (Miyashita & Russell, 1994). The individual *zaibatsu* had a monopoly in one or two industries, but soon the whole Japanese economy was divided between them. By the end of the First World War, each *zaibatsu* had launched at least one major manufacturing company in each sector and controlled, respectively, a bank, an insurance company, a shipping line, and a trading company. In 1930, approximately 75% of Japan's gross domestic product was directly or indirectly controlled by the largest *zaibatsu* (Kensy, 2001). After Japan's collapse in the Second World War, the American occupational forces decided to dissolve the *zaibatsu* as the source of Japanese military power. They intended to destroy the economic base of the Japanese military and prevent monopolistic market concentrations. As a result, it was planned to sell shares to the public and dissolve the *zaibatsu* into countless smaller companies (Baum, 1994). In 1947, the Antimonopoly Law came into effect. The new law made holding companies illegal (Miyashita & Russell, 1994). In 1948, the global political situation began to change. In

Europe, the Cold War started, and Communism began to spread in Europe and Asia. By 1948, the United States began to see Japan as a strategic buffer between the United States and the Communist countries. The United States needed a strong Japan with a strong economy rather than a weak Japan (Miyashita & Russell, 1994). The process of the dissolution of the *zaibatsu* was stopped. Many of them were re-established. The originally forbidden practice of using the old *zaibatsu* names in the names of new companies was now accepted (Morikawa, 1992). This time, companies grouped around the banks were allowed to hold shares in other companies, which made the establishment of financial links easier. They quickly achieved economic parity with the classic *zaibatsu*. These conglomerates were now called *keiretsu*. Some emerged out of former *zaibatsu*, but others were just new groupings of companies (Baum, 1994). The new *keiretsu* companies lacked parent corporations operating as holding companies; the influence of family, common in a *zaibatsu*, disappeared, and member companies were independent (Katsuki & Lennerfors, 2013).

References

Ahmadin, Ch. & Lincoln, E. J. (2001). Keiretsu, governance, and learning: Case studies in change from the japanese automotive industry. *Organization Science*, 12(6), 683–701.

Baum, J. A. C. (1994). Evolution of Keiretsu and their different forms. www.rotman.utoronto.ca/~baum/mgt2005/keiretsu.htm (Accessed May 10, 2024).

Freitag, M. (2004). Toyota. Formel Toyota. *Manager Magazin*, 12, 12–14.

Gabrowiecki, J. (2006). Keiretsu groups: Their role in the Japanese economy and a reference point (or Paradigm) for other countries. 413, 1–85.

Hoshi, T. (1994). The economic role of corporate grouping and the main bank system. In M. Aoki & R. Dore (eds.), *The Japanese Firm: The Sources of Competitive Strength* (pp. 285–309). Oxford: Oxford University Press.

Imai, M. (1986). *Kaizen. Der Schlüssel zum Erfolg der Japaner im Wettbewerb.* Frankfurt: Ullstein.

Katsuki, A. & Lennerfors, T. T. (2013). Organizational restructuring. The new, improved Keiretsu. In *Harvard Business Review*. https://hbr.org/2013/09/the-new-improved-keiretsu (Accessed May 10, 2024).

Kensy, R. (2001). *Keiretsu Economy — New Economy? Japan's Multinational Enterprises from a Postmodern Perspective.* New York: Palgrave.

Miyashita, K. & Russell, D. W. (1994). *Keiretsu: Inside the Hidden Japanese Conglomerates.* New York: McGraw Hill.

Morikawa, H. (1992). *Zaibatsu: The Rise and Fall of Family Enterprises Groups in Japan.* Tokyo: University of Tokyo Press.

Shimotani, M. (1995). The formation of distribution Keiretsu: The case of Matsushita electric. In E. Abe & R. Fitzgerald (eds.), *The Origins of Japanese Industrial Power: Strategy, Institutions and the Development of Organisational Capability* (pp. 54–69). London: Frank Cass.

Shun Cai, M., Goh, R., & Souza, G. (2013). Knowledge sharing in collaborative supply chains: Twin effects of trust and power. *International Journal of Production Research, 51*(7), 2060–2076.

Chapter 15

Lean Innovation — Kaikaku, Kaizen, and Innovation

15.1 Distinction between Kaikaku (改革) and Kaizen (改善)

Kaizen's philosophy is continuous improvement in small steps. But in reality, improvement is not uniform, but discontinuous, sometimes faster, and sometimes slower. With Kaizen, we sometimes face major obstacles and get stuck. In this case, we have to change the system, the framework conditions, in order to be able to improve further. A concentrated effort is required to make a breakthrough for Kaizen. This method is referred to as Kaikaku and helps to improve quality in production (Japanese: 改革, reformation, or transformation). Kaikaku is called "breakthrough kaizen" or "kaizen lightning." Kaikaku means "revolution in thought and action" and "improvement of great significance" (Radenkovic *et al.*, 2013). With a concentrated use of force, it temporarily accelerates the speed and scope of Kaizen. Kaizen is the continuous improvement of an activity with the purpose of increasing the performance of the production system, usually less than 20% or 30% in a given period of time (Radharmanan *et al.*, 1996). The opposite of Kaizen is Kaikaku, which represents drastic change or radical improvement and is usually a result of big investments in technology and/or equipment (Imai, 1986; Figure 15.1). This way of making improvements can be seen as a hindrance to constantly improving the process; thus, there is a need to conduct radical change in a company (Pfeffer, 1998). Kaikaku is not like Kaizen, which is started by all the employees (Radenkovic, 2013). Kaikaku usually starts with

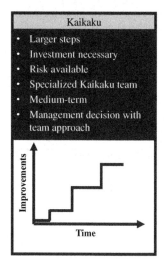 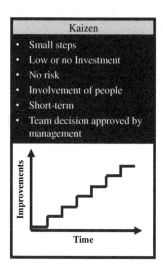

Figure 15.1.   Kaikaku and Kaizen.

the top management of the company and then goes to the lower levels of workers (Yamamoto, 2013). This is because Kaikaku represents crucial strategic changes. Table 15.1 compares Kaizen and Kaikaku (Helmold, 2021). Figure 15.1 shows the differences between the two methods.

15.2 Kaizen — Improvements in Small Steps

Kaizen (改善) is a Japanese management concept and targets improvements in small steps. Kaizen means all personnel are expected to stop their work when they encounter any abnormality and, along with their supervisor, suggest an improvement to resolve the abnormality. Kaizen aims to improve the quality of daily life, not just during working hours. The improvement should be gradual and infinite. It should pursue perfection (Helmold & Terry, 2021). The employees should be continuously engaged in the company's life and the improvement of every aspect of the company (processes, products, infrastructure, etc.) (Bertagnolli, 2020). This improvement throughout all aspects of life is related to the attention that is paid to the needs and requirements of customers (Nakano, 2020).

Kaizen focuses on teams (quality circles), promotes teamwork and team spirit; however, it also recognizes the individual contribution. It emphasizes the engagement of each worker in the concept and vision

Table 15.1.   Approach towards Kaizen and Kaikaku.

Kaizen (改善)	Kaikaku (改革)
Kaizen is based on multiple teams on the shopfloor/teams are guided by internal/external lean experts	Kaikaku is based on selected projects determined by top management/experts facilitate the projects
Kaizen projects are carried out throughout the value chain/teams assemble themselves	Kaikaku is focused on internal and external value chain processes/team is selected based on experience and capability

of the company, so that employees will identify themselves with the enterprise, its culture, and its objectives. The important aspects of Kaizen are:

- "What is wrong?" Not "who is wrong?"
- How do we eliminate waste (Muda)?
- How do we decrease quality costs?

Kaizen is a Japanese term and can be translated as "change for the better." The main goal of Kaizen is to continuously improve working areas, processes, and products by integrating the people of the affected areas. Usually, Kaizen is realized through workshops. Their typical duration can vary from three to five days. The aim of a Kaizen workshop is to implement the improvements during the time of the workshop. Kaizen is a framework combining the change of the company culture with the daily implementation of the principles (Liker, 2020). The 10 principles of Kaizen can be described as follows:

(1) Say no to the *status quo.*
(2) If something is wrong, correct it.

(3) Accept no excuses and make things happen.
(4) Improve everything continuously.
(5) Abolish old, traditional concepts.
(6) Be economical.
(7) Empower everyone to take part in problem-solving.
(8) Before making decisions, ask "why" five times to get to the root cause.
(9) Get information and opinions from multiple people.
(10) Remember that improvement has no limit, so never stop trying to improve.

A useful tool in the context of Kaizen is the P-D-C-A cycle. PDCA is an iterative four-step management method used in business for the control and continuous improvement of processes and products. It is also known as the Deming circle/cycle/wheel, or the Shewhart cycle. Since the 1950s, the PDCA cycle has been recognized as a simplified illustration of the elementary steps of a continuous improvement process:

Plan: Analyze the current situation and define an improvement plan.
Do: Implement the defined solutions.
Check: Evaluate the improvement results.
Act: Define counteractions in case of deviation from objective, standardize the best solution.

After improvement, it is important to standardize and implement the action so that the process or activity cannot return to its old state. If this is secured, one can aim for the next improvement.

15.3 Kaizen in Comparison to Innovation

Kaizen is the concept of small improvements in small steps, as shown in Figure 15.2 (Ohno, 1990). In contrast to an innovation, which is a top-down-approach, Kaizen involves all team members (see Figure 15.3). It means improvement and continuing to improve in personal life, home life, social life, and working life. When applied to the workplace, this philosophy means continuous improvement involving everyone, i.e., managers and workers alike (Kaizen Institute, 2019). The principles of Kaizen are customer knowledge and transparency. Thus, it is possible to improve a process without major investments. Kaizen in any organization is

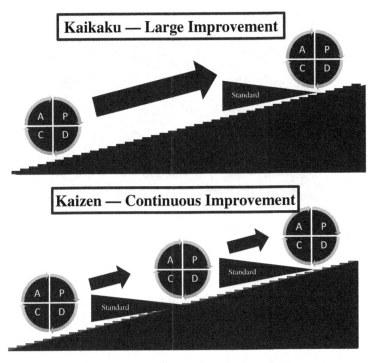

Figure 15.2.   Kaikaku (large improvement) and Kaizen (continuous improvement).

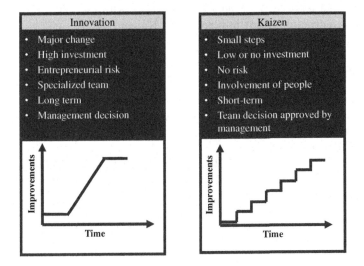

Figure 15.3.   Innovation and Kaizen.

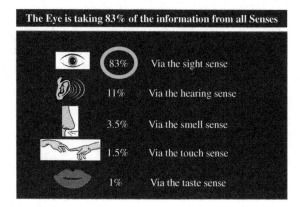

Figure 15.4. Visualization.

fundamentally important for a successful and continuous improvement culture and to mark a turning point in the progression of quality, productivity, and labor-management relations (Kaizen Institute, 2019).

15.4 Visualization Management

Visualization management is a significant part of Kaizen. Figure 15.4 displays that 83% of the issues are perceived with the eyes; thus, visualization is a crucial part of implementing Kaizen.

References

Bertagnolli, F. (2020). *Lean Management*. Wiesbaden: Springer.
Helmold, M. (2021). *Kaizen, Lean Management und Digitalisierung. Mit den japanischen Konzepten Wettbewerbsvorteile für das Unternehmen erzielen.* Heidelberg: Springer.
Helmold, M. & Terry, B. (2021). *Operations and Supply Management 4.0. Industry Insights, Case Studies and Best Practices.* Heidelberg: Springer.
Liker, J. K. (2020). *The Toyota Way. The Toyota Way: 14 Management Principles from the World's Greatest Manufacturer* (2nd edn.). Madison: Mc Graw-Hill.
Nakano, M. (2020). *Supply Chain Management. Strategy and Organization.* Cham: Springer.
Ohno, T. (1990). *Toyota Production System. Beyond Large Scale Production.* New York: Productivity Press.

Pfeffer, J. (1998). *The Human Equation*. Harvard Business School Press.

Radenkovic, M. *et al.* (2013). Improvement of quality in production progress by applying Kaikaku method. *International Journal for Quality Research*, 7(4).

Radharmanan, R., Godoy, L. P., & Watanabe, K. I. (1996). Quality and productivity improvement in a custom-made furniture industry using kaizen. *Computers & Industrial Engineering*, 3(1–2), 471–474.

Yamamoto, Y. (2013). *Kaikaku in Production. School of Innovation, Design and Engineering*, Malardalen University Press Licentiate Theses, Sweden, No. 120.

Chapter 16

Smart and Digital Total Productive Maintenance (TPM)

16.1 Scope and Definition

Total productive maintenance (TPM) started as a method of physical asset management and focused on maintaining and improving production machinery in order to reduce the operating costs of an organization. After the PM award was created and awarded to Nippon Denso in 1971, the Japanese Institute of Plant Maintenance (JIPM) expanded it to include eight activities of TPM that required participation from all areas of manufacturing and non-manufacturing in the concepts of lean manufacturing. TPM is designed to disseminate the responsibility for maintenance and machine performance, improving employee engagement and teamwork within management, engineering, maintenance, and operations.

- There are eight types of activities in the TPM implementation process:
  - Kobetsu-Kaizen (focused improvement) activities.
  - Jishu-Hozen (autonomous maintenance activity).
  - Planned Maintenance activity.
  - Hinshitsu-Hozen activity (quality maintenance activity).
  - Development Management activity.
  - Education and Training activity.

○ OTPM (office total productive maintenance, or office TPM).
○ Safety, Health, and Environment Activity.

16.2 Goal of TPM

TPM is an element in Lean Management (Bertagnolli, 2020). The goal of TPM is to improve equipment effectiveness by engaging those who impact it in small group-improvement activities. Total quality management (TQM) and TPM are considered the key operational activities of the quality management system. In order for TPM to be effective, the full participation of the entire organization, from top to frontline operators, is vital (Nakano, 2020). This should result in accomplishing the goal of TPM: Enhance the volume of production, employee morale, and job satisfaction.

The main objective of TPM is to increase the overall equipment effectiveness (OEE) of plant equipment. TPM addresses the causes of accelerated deterioration and production losses while creating the correct environment between operators and equipment to create ownership (Ohno, 1990).

OEE has three factors which are multiplied to give one measure called OEE: Performance × Availability × Quality = OEE.

Each factor has two associated losses, making six in total. These six losses are as follows:

- **Performance** = (1) Running at reduced speed – (2) Minor stops,
- **Availability** = (3) Breakdowns – (4) Product changeover,
- **Quality** = (5) Startup rejects – (6) Running rejects.

The objective finally is to identify, prioritize, and eliminate the causes of the losses. This is done by self-managing teams that solve problems. Employing consultants to create this culture is a common practice (Helmold & Terry, 2021).

16.3 Eight Pillars of TPM

The eight pillars of TPM (Figure 16.1) are mostly focused on proactive and preventive techniques for improving equipment reliability (Helmold, 2021):

(1) Autonomous maintenance — Operators who use all of their senses to help identify causes of losses.

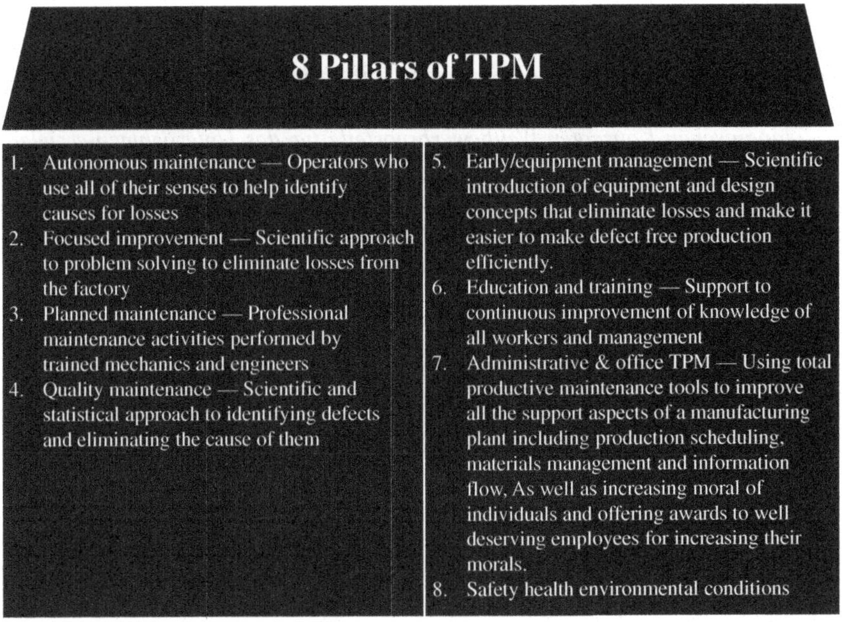

8 Pillars of TPM

1. Autonomous maintenance — Operators who use all of their senses to help identify causes for losses
2. Focused improvement — Scientific approach to problem solving to eliminate losses from the factory
3. Planned maintenance — Professional maintenance activities performed by trained mechanics and engineers
4. Quality maintenance — Scientific and statistical approach to identifying defects and eliminating the cause of them

5. Early/equipment management — Scientific introduction of equipment and design concepts that eliminate losses and make it easier to make defect free production efficiently.
6. Education and training — Support to continuous improvement of knowledge of all workers and management
7. Administrative & office TPM — Using total productive maintenance tools to improve all the support aspects of a manufacturing plant including production scheduling, materials management and information flow, As well as increasing moral of individuals and offering awards to well deserving employees for increasing their morals.
8. Safety health environmental conditions

Figure 16.1.   Pillars of TPM.

(2) Focused improvement — Scientific approach to problem-solving to eliminate losses from the factory.

(3) Planned maintenance — Professional maintenance activities performed by trained mechanics and engineers.

(4) Quality maintenance — Scientific and statistical approach to identify defects and eliminate their causes.

(5) Early/equipment management — Scientific introduction of equipment and design concepts that eliminate losses and make it easier to make defect-free production efficiently.

(6) Education and training — Supporting continuous improvement of knowledge of all workers and management.

(7) Administrative & office TPM — Using TPM tools to improve all the support aspects of a manufacturing plant, including production scheduling, materials management, and information flow, as well as boost the morale of individuals and offer awards to well-deserving employees for increasing their morale.

(8) Safe health environmental conditions.

References

Bertagnolli, F. (2020). *Lean Management*. Wiesbaden: Springer.

Helmold, M. (2021). *Kaizen, Lean Management und Digitalisierung. Mit den japanischen Konzepten Wettbewerbsvorteile für das Unternehmen erzielen.* Heidelberg: Springer.

Helmold, M. & Terry, B. (2021). *Operations and Supply Management 4.0. Industry Insights, Case Studies and Best Practices.* Heidelberg: Springer.

Nakano, M. (2020). *Supply Chain Management. Strategy and Organization.* Cham: Springer.

Ohno, T. (1990). *Toyota Production System. Beyond Large Scale Production.* New York: Productivity Press.

Chapter 17

Smart and Lean Factory

17.1 Scope and Definition

Smart factory, or intelligent factory, is a term from operations management and lean management used in the field of digitalized manufacturing technology. It is part of the German federal government's high-tech strategy for the future Industry 4.0 project. It describes the vision of a production environment in which manufacturing plants and logistics systems largely organize themselves without human intervention in order to produce the desired products (Helmold, 2024). A smart factory is a manufacturing solution that provides flexibility with lower costs and adaptive production processes, solving some of the problems that arise at production facilities in a world of increasing complexity. This manufacturing solution is based on automation, a combination of software, hardware, and mechanics, which leads to optimization of production processes, and reducing unnecessary labor and wastage of resources. The smart factories are pioneering examples of a new era of lean and digital production methods. They are the culmination of the digital era of the last 20 years, meeting the age-old practice of manufacturing. Smart factories illustrate the growing necessity for smarter production processes, smarter products, and hopefully an even smarter future. Figure 17.1 shows the smart factory elements with intelligent robots, automatic guided vehicles (AGVs), the Internet of Things (IOT), additive and flexible manufacturing, integrated supply chains, smart maintenance, big data, predictive analytics, realities, and cyber-security in planning, production, and after-service across all value chain stages.

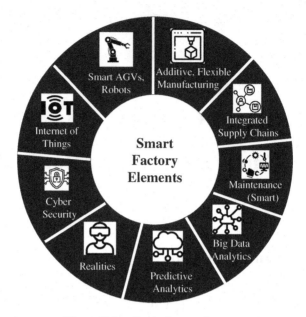

Figure 17.1.   Smart factory elements.

17.2 Digital Twin

A digital twin is a digital representation of a physical object, person, or process, contextualized in a digital version of its environment. Digital twins can help an organization simulate real situations and their outcomes, ultimately allowing it to make better decisions (Stjepandic *et al.*, 2023). In a smart factory, machines, tools, resources, products, and storage systems receive a digital representation, a so-called digital twin. The condition of all components can be viewed in real time at any time via the digital twin representations. So-called cyber-physical systems (CPS) are created from mechanical and electronic components that are wirelessly networked with one another and communicate in the "IIoT" or the Industrial IOT.

This communication is ensured by built-in chips in workpieces and production systems that are connected to the internet for data exchange. Workpieces to be processed become smart products by providing data about their own manufacturing process during manufacturing, e.g., data saved on an radio-frequency identification (RFID) chip. These data are used to control the path of the smart products through the system and individual production steps. Thanks to the mass of data, production systems learn to make decisions independently with artificial intelligence (AI), react to unforeseen events, and solve complex tasks. In this scenario, data have become more of an economic asset and competitive advantage than ever before. The aim is to use the collected data and findings in such a way that, above all, the individual production of individual items tailored to customers is not only economically efficient and profitable but also mass customization, for example customer-specific mass customization production, is made possible (Rückert *et al.*, 2023).

17.3 Smart Factory Process

The smart factory process is described in Figure 17.2. The process integrates all participants in the value chain, from suppliers to end-customers. In smart factories, suppliers are connected with planning, production, and other processes through cloud systems and enterprise resource planning systems (ERPs), so that suppliers can anticipate and ideally balance the production with the actual customer needs. Logistics processes and partners are linked through real-time tracking and location information. It is

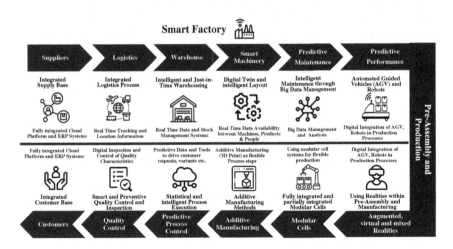

Figure 17.2. Smart factory process.

necessary to connect machines, products, and people with real-time data. Big data management supports these processes. Within the smart factory, robots and AGVs are used as part of the production process. Realities (augmented, virtual, and mixed), additive manufacturing methods, and integrated cells are used to optimize processes and production steps. Smart quality management tools and statistical process control enrich the process up to customer delivery to secure customer satisfaction (Helmold, 2024).

17.4 Areas of using Smart Factory Methods

17.4.1 Examples and Areas

The networking of embedded and production systems and dynamic business and engineering processes enable profitable production of products even for individual customer requirements up to batch size one. The technical basis is CPSs, which means both physical production objects and their virtual image are in a centralized system (see also digital twin) (Kletti & Rieger, 2023). In a broader context, the IOT is often mentioned. Part of this future scenario continues to be communication between the product (e.g., workpiece) and the production system: The product itself brings its production information in a machine-readable form, e.g., on an RFID chip (Budde *et al.*, 2023). These data are used to control the product's path through the production system and the individual production steps. Other transmission technologies, such as WLAN, Bluetooth, color coding, or QR codes, are currently being experimented with. At universities and research institutions, work is being done on the smart factory as part of so-called model factories, such as ARENA2036. Examples from industry include automobile production. For example, "Factory 56" from the Mercedes-Benz Group is equipped with WiFi and 5G networks. Various software applications network the machines and systems and enable big data analyses. The automobile manufacturer produces the Mercedes-Benz S-Class model in this factory and, from 2021, the electric variant Mercedes-Benz EQS (Günnel, 2020). In connection with the smart factory, similar concepts are also being worked on in areas that are directly linked to the production of products. For example, systems are being developed, especially in transportation, internal logistics, and material flow, that can move and transport the material autonomously, so that, for example, production can be supplied autonomously. Communication and location and identification technologies such as the Internet, WLAN play a role here; GPS or RFID play an important role.

17.4.2 The Tesla Gigafactory in Grünheide near Berlin

A better visualization of smart factories can be found in the example of Tesla's Gigafactory at Grünheide, Berlin (Figure 17.3). This is the fourth Gigafactory under construction, following the ones built in Nevada, Buffalo, and Shanghai. It also happens to be their first Gigafactory in Europe and could be remain the largest factory for many years. The Berlin factory was launched in 2021, and was the first factory of Tesla in Europe, with Tesla promoting it as the most advanced high-volume electric vehicle production plant in the world. It is used for building batteries, powertrains, and vehicles, starting with Model Y and Model 3, on a 300-hectare property in Grünheide, aiming at achieving a future production volume of 500,000 units annually. Tesla is striving to not only create a car that is smart but also to use smart means of production. The initial pictures of the factory released by the company's CEO, Elon Musk, depict their vision of a smart factory with solar panels dominating the rooftop, creating a more

Figure 17.3.   Tesla Gigafactory, Grünheide.

Source: https://www.automobilwoche.de/agenturmeldungen/tesla-will-mehrere-modelle-gigafactory-berlin-fertigen.

sustainable means of production. Its complex systems of producing cars have made headlines in the past, but the Berlin Gigafactory will take that up a notch. On its official website, Tesla mentions the use of a new dimension of casting systems and an efficient body shop, pushing forward the boundaries of vehicle safety. The paint shop introduces a new generation of color tone depth and complexity. Tesla's constant drive to transform how factories function has allowed it to revolutionize the car industry.

17.4.3 Adidas Speed Factory in Ansbach

The desire to create the factory of the future led Adidas to build a speed factory in Ansbach, Germany. With robots assisting humans, the company focused on sneaker-production processes in a single space, thereby completing production in a couple of days and proving their speed theory. Focusing on mass customization with shorter lead times, they use 3D printing technology to easily create digital replicas or mock-ups. Prototypes are quickly printed, facilitating an appropriate response to shifting consumer demands and satisfying customers' needs within days. Adidas may have shut down in Ansbach, yet they continue to use the speed factory technologies in the production of their athletic shoes in Asia. The company claims to have reinvented manufacturing in the footwear industry by replacing human hands in countries such as Indonesia, Vietnam, and China, by adopting technologies like 3D printing, robotic arms, multilayered particle machines, laser-cutting robots, and computerized knitting to make running shoes. They now aim to explore the use of 4D technology and other options to modernize their production. Figure 17.4 shows the Adidas factory.

17.4.4 The Connected Factory of Bosch in Blaichach

Global manufacturing of automotive technologies such as anti-lock braking systems (ABS) and electronic stability programs (ESP) requires efficiency in production. Bosch, at their lead plant in Blaichach (Figure 17.5), is doing exactly that. Implementing cutting-edge technology of connected manufacturing, the company uses 20 tablets every day to track processes, inspect machines, and check parts, data which are overlooked by 150 machine operators at the plant. Bosch's performance tracker system detects cycle-time deviations of mere milliseconds, enabling operators to

Figure 17.4.   Adidas Speed Factory in Ansbach.

Source: https://www.adidas-group.com/de/medien/newsarchiv/pressemitteilungen/2016/adidas-erweitert-mit-speedfactory-produktionskapazitaten-deutsuh/.

Figure 17.5.   The Connected Factory, Bosch, Blaichach.

Source: Bosch, https://www.bosch-presse.de/pressportal/de/de/bosch-stellt-weichen-fuer-die-fabrik-der-zukunft-195520.html.

react quickly and intervene at the earliest stage possible. Furthermore, an operator support system displays errors and provides recommendations on how to fix them (Bosch, 2019). Yet another technology is Bosch's Nexeed, a smart software suite that helps ensure production is running smoothly at their facilities. The company's connected industry cluster developed this software, which reads out data from 60,000 sensors and timely delivers the relevant information to the appropriate employee. The team thereby keeps track of manufacturing operations in real time, making predictive maintenance possible. Bosch's worldwide manufacture of ABS/ESP safety systems has seen a productivity jump of 24%, thanks to such skillful innovations.

17.4.5   Infineon's Smart Factory in Dresden

The smart factory set-up of Infineon in Dresden (Figure 17.6) wows the world with its intelligent networked manufacturing. The plant has achieved an automation level of 92%, with over 200 robots assisting employees. Infineon manufactures over 400 different products based on 200 mm and 300 mm wafers for all four of the group's segments, both quickly and in excellent quality. The 200 mm line is the world's most

Figure 17.6.   Infineon's Smart Factory in Dresden.
Source: https://www.infineon.com/cms/en/about-infineon/make-iot-work/smart-factory/.

automated factory, while its 300 mm line was set up to achieve fully automated production, improving its productivity by 70%.

The company's system controls, wafer transport, and production management are linked to each other and controlled in real time via IT systems. Furthermore, the systems can communicate with other Infineon sites around the world. Transport of the wafers is performed automatically, and the smart factory is controlled via central operating and monitoring systems as well as algorithms. The company optimizes production control by simulating the impact of changes to its product portfolio in advance. This gives them the freedom to flexibly respond to their customers' needs. Infineon continues to develop new products and solutions for the automotive and power electronics industries while investing in the latest technologies, such as AI, receiving attraction for all the right reasons. Figure 17.6 shows the smart factory of Infineon in Dresden (Infineon, 2023).

17.5 Challenge of Integration of Suppliers into the Smart Factory System

Specialized and global supply networks create complex supply networks. The challenge here is to integrate the supply base smoothly and smartly into the own value chain through intelligent and digital applications and ERPs. A method of short interval technology (SIT) can be used to systematically set up fast control loops in the company including stakeholders and suppliers. The goal is perfect production with transparent, responsive, and economical processes (Kletti & Rieger, 2023). With the classic methods of production planning and control as well as the frequently encountered production processes, it is only with great effort that many companies are able to operate in the increasingly better networked and faster-paced supply chains and achieve short delivery times, high adherence to delivery dates, small batch sizes, short-term call-offs, and just-in-time concepts that ensure on-time deliveries at ideal and competitive costs.

17.6 Advantages of Smart Factories

It is not a surprise that the smart factory is also called the factory of the future. Digital change has and will continue to have a major impact on industry and society in general. The smart factory model enables companies to keep up with change and implement customers' new needs for individuality, flexibility, and speed. The advantages of using smart factories include:

- Lean and optimized processes with shorter production times and lower costs, which lead to increased productivity.
- Customer-oriented, individual production of individual pieces up to batch size one at the cost of mass production.
- Automated and efficient ordering processes and supply chains by using predictive analytics or big data.
- Fast and agile adaptation to market requirements based on real-time information.
- Intelligent and transparent supply chain integration from suppliers to customers.
- More flexibility throughout the value chain with flexible manufacturing cells or additive manufacturing.
- Fast reactions to new or changed products and market requirements.
- Smart, digital, and preventive total productive maintenance process throughout the value chain.

References

Bosch. (2019). Industrie 4.0. Smart factory. Bosch stellt Weichen für die Fabrik der Zukunft. https://www.bosch-presse.de/pressportal/de/de/bosch-stellt-weichen-fuer-die-fabrik-der-zukunft-195520.html (Accessed January 4, 2024).

Budde, L. *et al.* (2023). *Smart Factory Navigator. Identifying and Implementing the Most Beneficial Use Cases for Your Company — 44 Use Cases That Will Drive Your Operational Performance and Digital Service Business.* Heidelberg: Springer.

Günnel, T. (2020). Smarte Fabrik. Mercedes-Benz eröffnet S-Klasse-Fabrik „Factory 56". Automobil-Industrie. https://www.automobil-industrie.vogel.de/mercedes-benz-eroeffnet-s-klasse-fabrik-factory-56-a-960628/ (Accessed January 5, 2024).

Helmold, M. (2024). *Erfolgreiche Transformation zum digitalen Champion: Wettbewerbsvorteile durch Digitalisierung und Künstliche Intelligenz.* Wiesbaden: Springer.

Infineon. (2023). Unleashing the power of the IoT. One-stop product portfolio and made-to-measure services for a faster time-to-market. Smart Factory.

https://www.infineon.com/cms/de/about-infineon/make-iot-work/smart-factory/ (Accessed January 8, 2024).

Kletti, J. & Rieger, T. (2023). *Die perfekte Produktion: Manufacturing Excellence in der Smart Factory*. Heidelberg: Springer.

Rückert, T. *et al.* (2023). *Digital Twin Development: An Introduction to Simcenter Amesim*. Wiesbaden: Springer.

Stjepandic, J. *et al.* (2023). *DigiTwin: An Approach for Production Process Optimization in a Built Environment*. Wiesbaden: Springer.

Chapter 18

Lean Management Trends

18.1 Lean Management Evolvement and Roadmap

Lean management is steadily evolving throughout sectors, industries, and value chains. Automation has enabled efficiency and accuracy, allowing workers to focus on more value-added tasks. Notably, there has been a shift toward a customer-centric approach focused on quality, flexibility, and speed. Plus, the implementation of new technologies such as big data and artificial intelligence has further facilitated the improvement of lean manufacturing processes and the identification of new opportunities for optimization. The implementation of lean practices begins with the actions which directly and indirectly create value for the customer. Through ongoing testing, our employees learn and innovate their work for increasingly better quality and flow, less time and effort, and lower waste. Distilled into five rotational steps, the lean management roadmap aims for the following:

- Specify the value desired by the customers and the benefits they will get.
- Identify the value stream for each product providing that value and challenge all of the wasted steps (generally 9 out of 10) currently necessary to provide it.
- Make the product flow continuously through the remaining value-added steps.
- Introduce pull between all steps in the value chain where continuous flow is impossible.
- Manage toward perfection so that the number of steps and the amount of time and information needed to serve customers continually falls.

18.2   Lean Continuation

Lean management does not only reduce operational waste but also optimizes the workforce. Integrating lean with modern technology enhances productivity, such as predictive maintenance in smart factories. Lean management remains indispensable. Its compatibility with new technologies empowers businesses to iterate, solve problems, and adapt swiftly in a rapidly changing world. In the age of big data, lean ensures accurate monitoring and a deeper understanding of customer benefits.

A real and profitable lean organization understands customer value and focuses its key processes to continuously increase it. The ultimate goal is to provide perfect value to the customer through a perfect value creation process that has zero waste. An ideal way to introduce a practical view in terms of lean thinking is to deploy the seven manufacturing basics, a handpicked selection of core lean improvement tools, methodologies, and techniques to grasp control of any shop-floor situation. Lean Management is using a set of agile, lean and effective tools and methods for various cases and in various situations. But as time progressed, it became very apparent that some combinations of these lean tools and elements worked very well together when deployed in a specific sequence. They should be part of the skillset a team leader uses in daily work on Jidoka and Kaizen reinforcement, assuring quality, controlling processes, and making incremental improvements often.

- Autonomous Maintenance ensures that enterprises have the pre-conditions for production set-up before running the process; equipment can run at a rate without losses or interruptions.
- Gemba Cadence establishes a fixed routine for observing employees and processes at work in the real environment.
- Change Point Management captures and controls all unplanned change points. Unplanned events are the silent killers that induce waste, overall equipment effectiveness (OEE) losses, defects, and workarounds.
- Critical Thinking Mentality helps an employee analyze the root cause and implement improvements.
- See disconnects in the entire process — examine the four levels of how any process is configured: systems, pathways, connections, and activities — to confirm it is working as intended.

18.3 Synchronization of Value Chain Networks

Future cloud platforms and ERP systems will be integrated into the entire value chain, from raw material makers to the final end customers. Digitalization will enable a transparent and sustainable supply chain, including departments such as procurement, operations, marketing, planning, logistics, finance, human resources, and distribution (Helmold & Terry, 2021).

18.4 Flexible Sensors

Manufacturers want flexible processes that enable them to manufacture several products on one production line. The benefits of flexibility are difficult to grasp; however, time-consuming retooling is required to prepare machines to manufacture different products (Helmold & Terry, 2021). By implementing lean tools, such as the one-minute exchange of tools, manufacturers can remove non-value-adding activities from the conversion and thus speed up the process considerably. Companies that have implemented lean methods and Industry 4.0 can benefit from these technologies (Küpper *et al.*, 2017). New sensors and self-learning software enable machines to automatically identify products and load the appropriate program and tools without manual intervention. As the changeover is automated, operators can focus on value-adding activities.

18.5 Transparency, Flexibility, and Agility

Lean management tools of the future (Lean Management 2030) will help companies make complex and global value streams transparent and manageable. Lean offers standardized tools and methods with which companies can concentrate on customer benefits and actively involve and continuously improve each individual employee (Bertagnolli, 2018). But what if people no longer have the opportunity to examine the complexity and thus continually make suggestions for improvement? Big data and additive manufacturing as exemplary drivers of digitization are also used in manufacturing companies and in sectors such as healthcare or service. A continuous introduction to other technology branches will be visible in the next 10 years. Manufacturers who have already successfully used Lean Industry 4.0 can reduce conversion costs by up to 40% in 5–10 years in order to be ready for Lean Management 2030, considerably better than the savings that were achieved through the first-class independent use of Lean Industry 4.0. In many cases, higher cost reductions are achieved with technologies that improve plant processes and structures, for example, by optimizing layouts. However, less than 5% of the manufacturing companies we observed have reached a high level of maturity in Lean Industry 4.0 (Siebenmorgen, 2016). To achieve the greatest benefits, a manufacturer must adapt the application of Lean Industry 4.0 to address its specific challenges along the supply chain and at the plant level. A recent global survey by the Boston Consulting Group found that leading industrial companies recognize the importance of lean management and digitization for their long-term planning. In a survey of more than 750 production managers, 97% of respondents in the automotive sector said that lean management would be highly relevant in 2030, compared to 70% who consider it important today. Among these respondents, 70% stated that the successful digitization will be highly relevant to the business success by 2030, whereby 13% stated that the digital transformation as relevant and important (Küpper *et al.*, 2017).

18.6 Implementation of intelligent Poka Yoke and Quality Systems

Digital quality systems and Poka-Yoke Lean Management 2030 methodologies will use digital error prevention appliances. Production capacity is wasted if products do not meet specifications. Even worse, if a company ships poor-quality products to customers, they will incur higher costs and likely lose trust in that supplier. Many lean management tools, such as self-inspection, *poka-yoke*, and *jidoka*, have been developed to reduce the probability of mistakes and increase the rate and speed of error detection. For example, the analysis by Boston Consulting Group shows that self-inspections improve the process of providing feedback to engineers and operators, thereby accelerating error detection and reducing the number of defects by 50%–70%. However, to achieve zero defects, manufacturers must support self-inspections by using a data-driven analytics approach to identify the root causes of errors. Industry 4.0 technologies and innovative lean methods allow such support by providing reliable context data and the ability to conduct detailed tracking. The analysis of errors is enhanced through, for example, camera-based visual inspection, correlation models, and real-time monitoring of process parameters (Küpper *et al.*, 2017).

18.7 Collaborative Robots (Cobots) as Supplementary Partners in Lean Production Environments

The automation industry has been discussing the vision of batch size one for a number of years. The question emerging with this issue is, how can production systems manufacture individual items without long retooling or becoming inefficient? With the possibilities of Industry 4.0 and innovative lean management tools, it will not be long before this vision becomes a reality and enables comprehensive customization in production. In order to achieve this goal, machines in production can no longer be used in a rigid and inflexible manner. So far, they have mostly only been put into operation, parameterized, and coordinated for a specific product, which is then manufactured over and over again over months or even years. Tomorrow's production lines must be flexible — built from several mechatronic modules that can easily be redesigned and recombined with more and more robots, or "cobots" (collaborative robots that work hand in hand with people). In addition, an AI should be integrated that parameterizes and adjusts the machines for the next — individualized — product that is to be manufactured.

18.8 Lean Management Will Include Algorithms and Virtual Tools

Lean management will include algorithms and virtual tools in 2030. In many manufacturing industries, device failures result in high inventory levels, significant working capital costs, and low efficiency (Küpper *et al.*, 2017). Organizations can use lean methods such as autonomous or preventive maintenance to dramatically increase OEE. For example, by using autonomous maintenance, companies put their operators in charge of certain do-it-yourself maintenance activities, which significantly reduces downtime for fixing minor issues. Leading manufacturers make the most of these lean methods by using advanced analysis algorithms and machine learning techniques to analyze the huge amounts of data that are being captured by sensors. The output identifies the potential for failures before they occur (Henao *et al.*, 2019). Such predictive insights prepare operators to carry out autonomous maintenance at the optimal point in time, which reduces disruptions and minimizes unnecessary downtimes and replacement costs.

18.9 Lean Management Enables an Ethical Compass

Lean progresses in the context of specific situations: respect for the humanity of customers, employees, suppliers, investors, and our communities, with the belief that all can and will be better off through lean practices (Helmold, 2021).

18.10 Improving Environmental Sustainability

Lean identifies environmental impacts, aiding in sustainable production. Lean principles, such as operator care, reduce waste and promote efficient resource use. Industry 4.0 lighthouse companies have successfully reduced emissions and improved resource management.

References

Helmold, M. (2021). *Kaizen, Lean Management und Digitalisierung. Mit den japanischen Konzepten Wettbewerbsvorteile für das Unternehmen erzielen.* Wiesbaden: Springer.

Helmold, M. & Terry, B. (2021). *Operations and Supply Management 4.0. Industry Insights, Case Studies and Best Practices.* Cham: Springer.

Henao, R., Sarache, W., & Gómez, I. (January 20, 2019). Lean manufacturing and sustainable performance: Trends and future challenges. *Journal of Cleaner Production*, 208, 99–116.

Küpper, D. *et al.* (2017). When lean meets Industry 4.0. The next level of operational excellence. Boston Consulting Group. https://www.bcg.com/publications/2017/lean-meets-industry-4.0.aspx (Accessed on November 28, 2019).

Siebenmorgen, F. (2016). Industrie 4.0. Das Potenzial schon heute nutzen. https://www.supplyon.com/wp-content/uploads/import/DE_SCM%20 Magazin_Industrie%204.0.pd (Accessed on November 28, 2019).

Index

Printed in the United States
by Baker & Taylor Publisher Services